Exam Secrets

AS

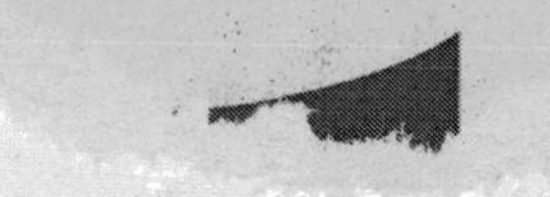

Biology

John Parker

Contents

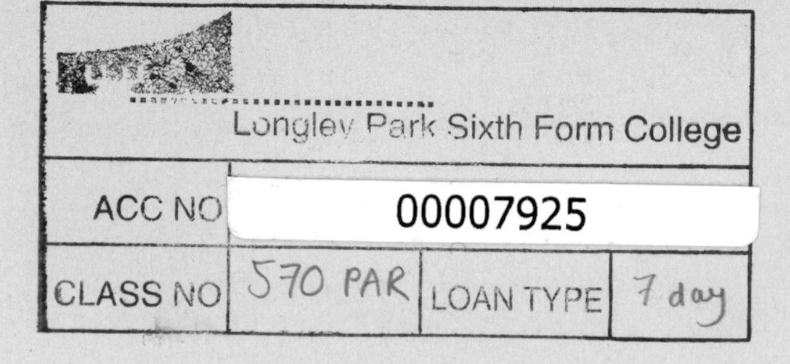

Examination boards

Each Examination Board can be contacted at the addresses and websites printed below. Each Board can supply specific information about assessment regulations and specification issues. If it is necessary to contact them, inform them of which specification you are studying. In the case of AQA, it may be either specification A or B. Your Exam Board can give more detailed information about topics covered and exams taken.

AQA Assessment and Qualifications Alliance
Devas Street,
Manchester,
M15 6EX
www.aqa.org.uk

EDEXCEL
Stewart House,
32 Russell Square,
London,
WC1B 5DN
www.edexcel.org.uk

OCR Oxford Cambridge and RSA Examinations
1 Hills Road,
Cambridge,
CB1 2EU
www.ocr.org.uk

CCEA Northern Ireland Council for Curriculum, Examinations and Assessment
29 Clarendon Road,
Belfast,
BT1 3BG
www.ccea.org.uk

WJEC Welsh Joint Education Committee
245 Western Avenue,
Cardiff,
CF5 2YX
www.wjec.co.uk

AS exams

Different types of examination questions

Structured questions

Structured questions are in several parts. Each will normally begin with a brief amount of stimulus material, in the form of a diagram, data or graph. The parts are usually about a common context and they often progress in difficulty. They may start with simple recall, then test understanding of a familiar or unfamiliar situation.

The most difficult part of a structured question is usually at the end. Ascending in difficulty, a question allows a candidate to build in confidence. Technological and social applications of biological principles give a more demanding challenge. Most of the questions in this book are structured questions. This is the main type of question used in the assessment of AS Biology.

Extended questions

In AS Biology questions requiring more extended answers usually form part of structured questions. These extended answers normally appear at the end of structured questions and typically have a value from four to twenty marks. These questions are allocated more lines, so you can use this as a guide as to how many points you need to make in your response. Typically for an answer worth ten marks the mark scheme would have around 12 → 14 creditable answers. You can score up to the maximum, ten marks. Extended answers are used to allocate marks for the **quality of communication**.

Candidates are assessed on their ability to use a **suitable style of writing, and organise relevant material, both logically and clearly**. The use of **specialist biological terms** in context is also assessed. **Spelling, punctuation and grammar** are also taken into consideration.

What examiners look for

Examiners are looking for logical answers to their questions so it is important to read the stimulus material carefully. Your answer will be credited if it contains the main facts. You do not get extra marks for writing a lot of additional words. Keep to the point! Make sure that your answer is clear, easy to read and concise.

What makes an A, C and E grade candidate?

This is useful information if you wish to achieve your best potential grade. The way to do this is to make sure that you have a good, all round, knowledge and understanding of Biology.

A grade candidates recall and use biological knowledge, facts, principles and concepts from the complete specification, displaying only minor gaps in knowledge and understanding; select biological knowledge for responses to questions, which is both relevant and logically presented; use a range of specific technical terms in context; carry out a range of calculations accurately, in a logical manner. The minimum mark for an A grade candidate is 80%.

C grade candidates recall and use biological knowledge, facts, principles and concepts reliably, from many parts of the specification; frequently select biological knowledge for responses to questions, which is in context and presented clearly and logically; frequently use appropriate technical terms; and carry out a limited range of calculations accurately. The minimum mark for a C grade candidate is 60%.

E grade candidates demonstrate limited understanding of biological knowledge, facts, principles and concepts from some parts of the specification; show an understanding of basic principles and concepts beyond that expected of sound GCSE candidates; use some basic technical terms across the specification and carry out a limited range of straightforward calculations. The minimum mark for an E grade candidate is 40%.

Successful revision

Revision skills

- Begin your revision programme several weeks before your examination.
- Count-down to your exam day – small amounts of revision increasing in time as you approach the examination will be beneficial.
- Writing down basic points on cards will help commit concepts to long-term memory – it is important to have a reservoir of knowledge from which to retrieve facts during your exam.
- Last minute exam-cram is of little use. You need to assimilate the information gradually and be able to apply the concepts to the new situations you will meet in the exam modules.
- Learn processes in sequence, carefully. Less able candidates miss out important parts of their answers and often corrupt the topics they have been taught. There are very few marks for being almost correct!
- Learning takes time. Success does not come easily. How hungry are you to earn your success?
- The most important part of your revision programme is to answer questions. The more you answer the better you will become.
- Without feedback, answering questions would be of no value. However, coupled with guidance from your teacher and this book, your strengths and weaknesses will be identified. Concentrate on weak area topics then see your performance improve … dramatically!

Practice questions

This book specialises in both questions and feedback. For the perfect solution use the *Letts Revise AS Study Guide* to support you. Their bullet-point style follows the way mark schemes are written. The main points regularly examined are included and published in an easy-to-learn format.

- Look at the answers of the A and C candidates. See if you could have done better.
- Check out the points missed by the C candidates as well as the major errors. This book includes errors regularly made by C, D and E candidates. Noting regular misconceptions will help you to avoid the same errors.
- Try to do the exam practice questions then check them out against the mark scheme (see page 6).
- Make sure that you understand why the answers given are correct.
- Going through the chapters should help you prepare for specific modules.

Use the book regularly and use a highlighter pen to remind you of key information.

Planning and timing your answers

- Scan each question before you begin to answer. The key information is the stimulus material at the beginning.
- Try to link the question with a specific part of the syllabus. You will then be able to remember the principles that are being tested. Remember that examiners often repeat similar questions.
- Check out the mark allocation on the right of each question. This shows how many points you need to make in your answer.
- Plan your answers. Writing the first thing that comes into your head gains little credit. Keep to the point. Never give extra information just because you think you know it!
- Present your answers clearly. There is nothing to gain by hurrying your responses so much that the examiner cannot understand your writing. Aim for clarity.

How to boost your grade

Learn all definitions throughout the specification. These are straightforward marks to gain, but you must prepare well. Along with labelling diagrams, they are among the easier marks to gain successfully.

Graphs appear in a lot of questions. Be ready to interpret the relationship shown by the data. If asked to **describe** the relationship shown, this may be simply a comment linking the two variables, *e.g. the amount of product produced in an enzyme catalysed reaction increases to a peak, then falls with temperature increase*.

Explain If the question requires you to **explain** the relationship then the reasons for the changes in data are required. In this example you would need to refer to the shape of the active site of the enzyme molecules and collisions between substrate and enzymes at different temperatures.

In explaining relationships, when more than one graph line is shown, look for peaks and troughs. Where a peak of one graph line corresponds with a trough of another then it is likely that they are linked, e.g. where pressure in a heart atrium is high, and the volume is reducing, then a student could conclude that the atrium was both contracting and emptying.

Where a change takes place, then you may be expected to think of a reason not shown on the graph, *e.g. the number of aphids increases dramatically from April to September. You may be expected to answer in terms of higher Spring temperatures. Greater plant growth in Spring should be related to additional food for the aphids so reproduction takes place.*

Detail is required at this level. The following example will illustrate this point. A question requires the candidate to explain how oxygen is passed to the tissues.

The candidate writes, The medulla is linked to the heart by nerves, which can speed up, or slow down the heart rate. Electrical stimulation causes the atria and ventricles to contract.

This answer lacks the required detail! This is the detail required.

The medulla oblongata is linked to the heart by the sympathetic nerve,which speeds up the heart rate;
and the vagas nerve which slows down the the heart rate;
At the tissues carbon dioxide is formed during respiration;
The nerves link to the sino-atrial node (SAN);
and from here a wave of electrical activity passes across both atria which contract;
The electrical activity reaches the atrio-ventricular node (AVN);
and from here electrical activity passes down the bundle of His;
This stimulates the ventricles to contract from the base upwards;

The first answer shows a superficial answer from a student around the E/U grade boundary. AS demands much more detail. Set out to learn all the details. Learn it all in logical steps.

This book shows the detail of official mark schemes. Go for ultimate detail and achieve the ultimate grade, A!

Revise You need to do **Regular Revision** through the course; this keeps the concepts 'hot' in your memory, 'simmering and distilling', ready to be **retrieved** and **applied** to new situations.

How to use the mark scheme

Symbol	**Meaning**
;	A separate mark
/	An alternative answer acceptable
max	Mark scheme shows that marks available exceed the question value You can be awarded up to the maximum
<u>underline</u>	When the word or phrase is underlined then it must be given if a mark is to be awarded

Chapter 1

Biological molecules

Questions with model answers

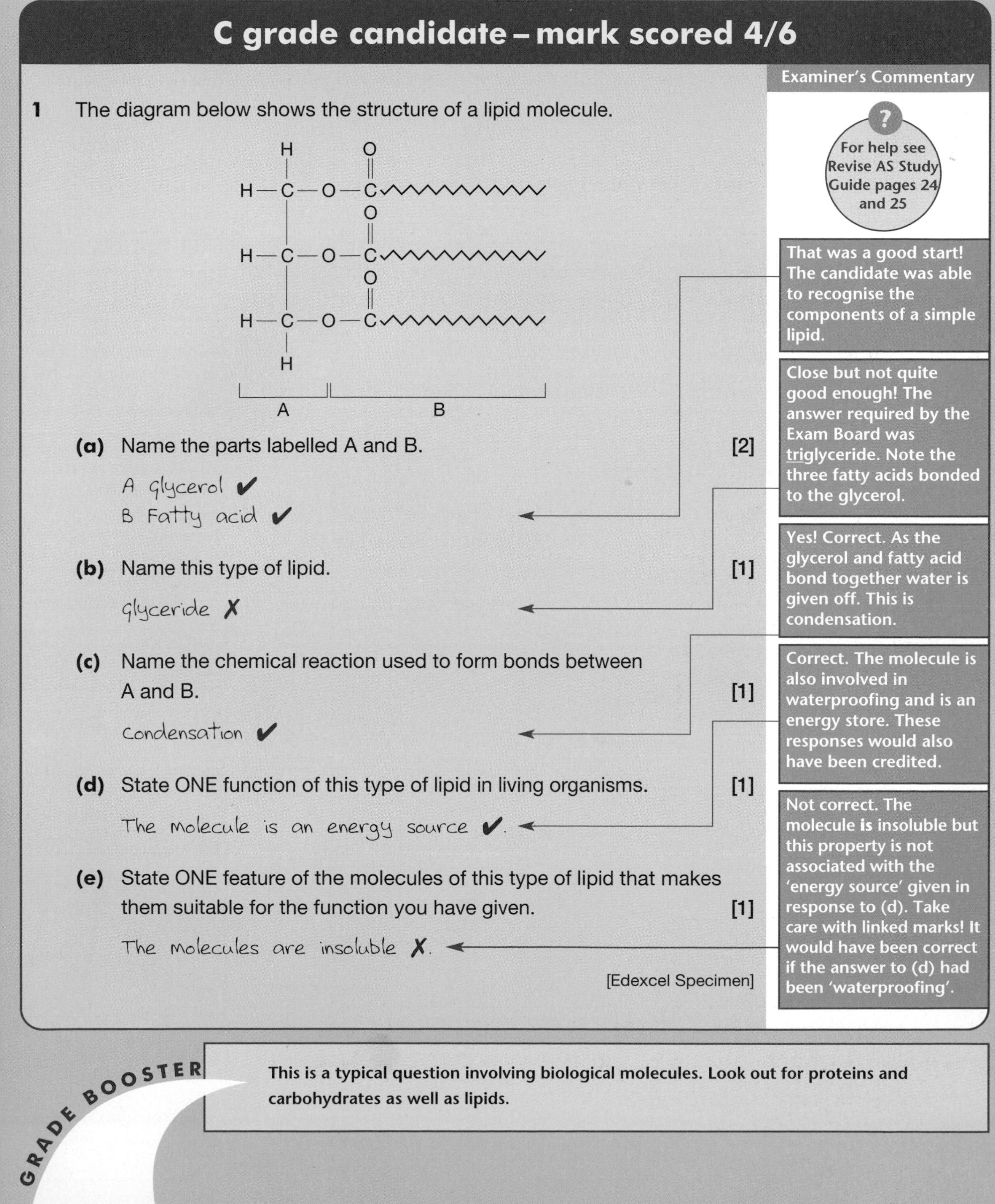

C grade candidate – mark scored 4/6

1 The diagram below shows the structure of a lipid molecule.

(a) Name the parts labelled A and B. **[2]**

A glycerol ✔
B Fatty acid ✔

(b) Name this type of lipid. **[1]**

Glyceride ✗

(c) Name the chemical reaction used to form bonds between A and B. **[1]**

Condensation ✔

(d) State ONE function of this type of lipid in living organisms. **[1]**

The molecule is an energy source ✔.

(e) State ONE feature of the molecules of this type of lipid that makes them suitable for the function you have given. **[1]**

The molecules are insoluble ✗.

[Edexcel Specimen]

Examiner's Commentary

For help see Revise AS Study Guide pages 24 and 25

That was a good start! The candidate was able to recognise the components of a simple lipid.

Close but not quite good enough! The answer required by the Exam Board was <u>tri</u>glyceride. Note the three fatty acids bonded to the glycerol.

Yes! Correct. As the glycerol and fatty acid bond together water is given off. This is condensation.

Correct. The molecule is also involved in waterproofing and is an energy store. These responses would also have been credited.

Not correct. The molecule **is** insoluble but this property is not associated with the 'energy source' given in response to (d). Take care with linked marks! It would have been correct if the answer to (d) had been 'waterproofing'.

GRADE BOOSTER

This is a typical question involving biological molecules. Look out for proteins and carbohydrates as well as lipids.

A grade candidate – mark scored 14/15

Examiner's Commentary

For help see Revise AS Study Guide pages 24 and 25

2 PrP is a protein normally found in brain tissue. The diagram shows part of the structure of a PrP molecule. The dotted line represents the middle part of the molecule which has been left out of this diagram.

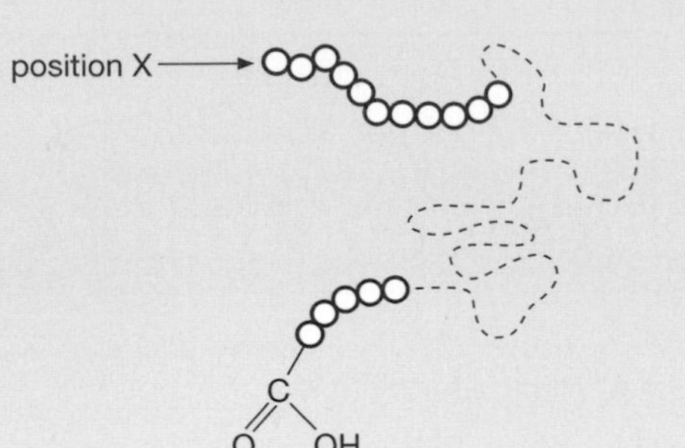

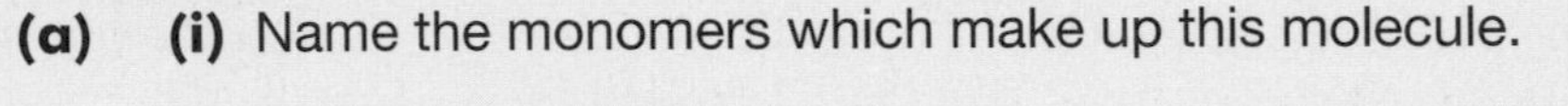

(a) **(i)** Name the monomers which make up this molecule. [1]

Amino acids ✔

> The key to scoring this mark is the – COOH (carboxylic acid group). A polymer with this group at one end points to a polypeptide. Polypeptides consist of linked amino acids.

(ii) Name ONE chemical element which would be found in this molecule but not in a polysaccharide molecule. [1]

Nitrogen ✔

> Correct! The exam board accepted sulphur from the sulphur bridges, although none are shown in the diagram.

(iii) Give the formula of the chemical group that would be found at position X on the molecule. [1]

Amine group ✗

> No! That was a slip up. The candidate probably knew the answer but failed to give the **formula**, NH_2. Even high ability candidates slip up sometimes but an A is still possible! You don't have to score 100% to be awarded an A.

PrP molecules are found on the outside of the cell surface membranes of nerve cells. The precise function of the PrP is still unknown but it is thought that its tertiary structure enables it to act as a receptor molecule.

(b) Describe:

(i) the secondary structure of a protein [1]

α helix or β pleated sheet ✔

> Correct. The candidate gave two correct answers, where only one would have been enough. Valuable time was lost.

(ii) the tertiary structure of a protein. [1]

Further folding of a polypeptide ✔.

> Correct. A globular shape would also have been credited, this being the result of the further folding already given by the candidate.

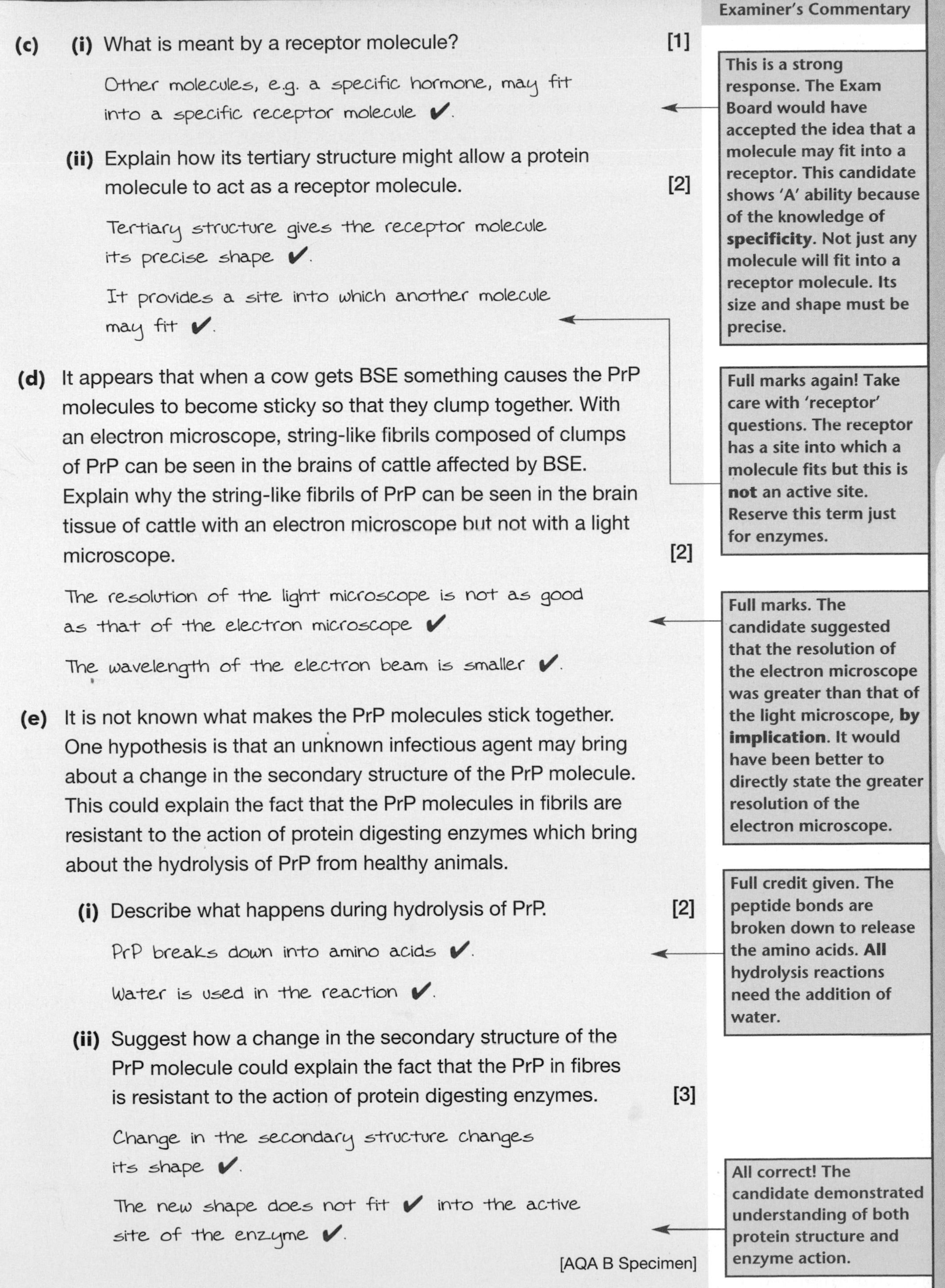

Examiner's Commentary

(c) (i) What is meant by a receptor molecule? [1]

Other molecules, e.g. a specific hormone, may fit into a specific receptor molecule ✔.

This is a strong response. The Exam Board would have accepted the idea that a molecule may fit into a receptor. This candidate shows 'A' ability because of the knowledge of **specificity**. Not just any molecule will fit into a receptor molecule. Its size and shape must be precise.

(ii) Explain how its tertiary structure might allow a protein molecule to act as a receptor molecule. [2]

Tertiary structure gives the receptor molecule its precise shape ✔.

It provides a site into which another molecule may fit ✔.

Full marks again! Take care with 'receptor' questions. The receptor has a site into which a molecule fits but this is **not** an active site. Reserve this term just for enzymes.

(d) It appears that when a cow gets BSE something causes the PrP molecules to become sticky so that they clump together. With an electron microscope, string-like fibrils composed of clumps of PrP can be seen in the brains of cattle affected by BSE. Explain why the string-like fibrils of PrP can be seen in the brain tissue of cattle with an electron microscope but not with a light microscope. [2]

The resolution of the light microscope is not as good as that of the electron microscope ✔

The wavelength of the electron beam is smaller ✔.

Full marks. The candidate suggested that the resolution of the electron microscope was greater than that of the light microscope, **by implication**. It would have been better to directly state the greater resolution of the electron microscope.

(e) It is not known what makes the PrP molecules stick together. One hypothesis is that an unknown infectious agent may bring about a change in the secondary structure of the PrP molecule. This could explain the fact that the PrP molecules in fibrils are resistant to the action of protein digesting enzymes which bring about the hydrolysis of PrP from healthy animals.

(i) Describe what happens during hydrolysis of PrP. [2]

PrP breaks down into amino acids ✔.

Water is used in the reaction ✔.

Full credit given. The peptide bonds are broken down to release the amino acids. **All** hydrolysis reactions need the addition of water.

(ii) Suggest how a change in the secondary structure of the PrP molecule could explain the fact that the PrP in fibres is resistant to the action of protein digesting enzymes. [3]

Change in the secondary structure changes its shape ✔.

The new shape does not fit ✔ into the active site of the enzyme ✔.

All correct! The candidate demonstrated understanding of both protein structure and enzyme action.

[AQA B Specimen]

Exam practice questions

1 The statements in the table refer to three polysaccharide molecules.
If the statement is correct place a tick (✔) in the appropriate box and if the statement is incorrect place a cross (✗) in the appropriate box.

Statement	Starch	Glycogen	Cellulose
Polymer of α glucose			
Glycosidic bonds present			
Unbranched chains only			
Energy store in animal cells			

[4]

[Edexcel Specimen]

2 The diagram shows how some biological molecules may be separated from each other.

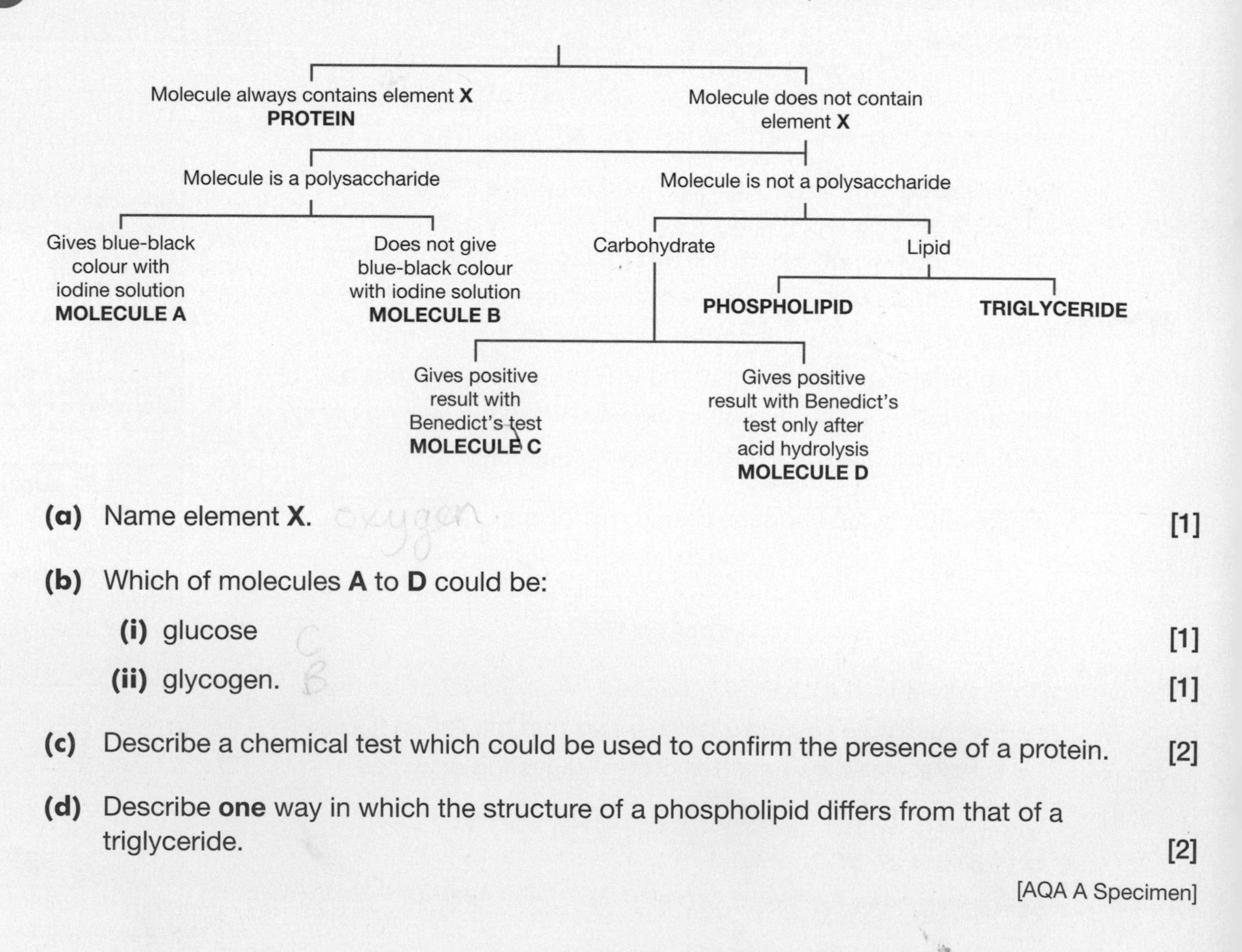

(a) Name element **X**. **[1]**

(b) Which of molecules **A** to **D** could be:

(i) glucose **[1]**

(ii) glycogen. **[1]**

(c) Describe a chemical test which could be used to confirm the presence of a protein. **[2]**

(d) Describe **one** way in which the structure of a phospholipid differs from that of a triglyceride. **[2]**

[AQA A Specimen]

Answers on pages 13–14 **Answers** on pages 13–14 **Answers** on pages 13–14

3 **(a)** For each of the following pairs of carbohydrates, give one word which describes **both** of the molecules and **distinguishes** them from the other pairs.

(i) Cellulose and glycogen **(ii)** Maltose and lactose

(iii) Ribose and deoxyribose **(iv)** Glucose and fructose [4]

(b) Name the reagent for testing for reducing sugars in food. [1]

(c) Name a sugar that would not give a positive test with this reagent. [1]

[WJEC Specimen]

---

4 A dipeptide was heated with an acid for 20 minutes. Three different samples were taken and loaded onto a piece of chromatography paper. These were:

1. the dipeptide only

2. the dipeptide after it had been heated with the acid for 2 minutes

3. the dipeptide after it had been heated with the acid for 20 minutes.

The resulting chromatogram is shown in the drawing.

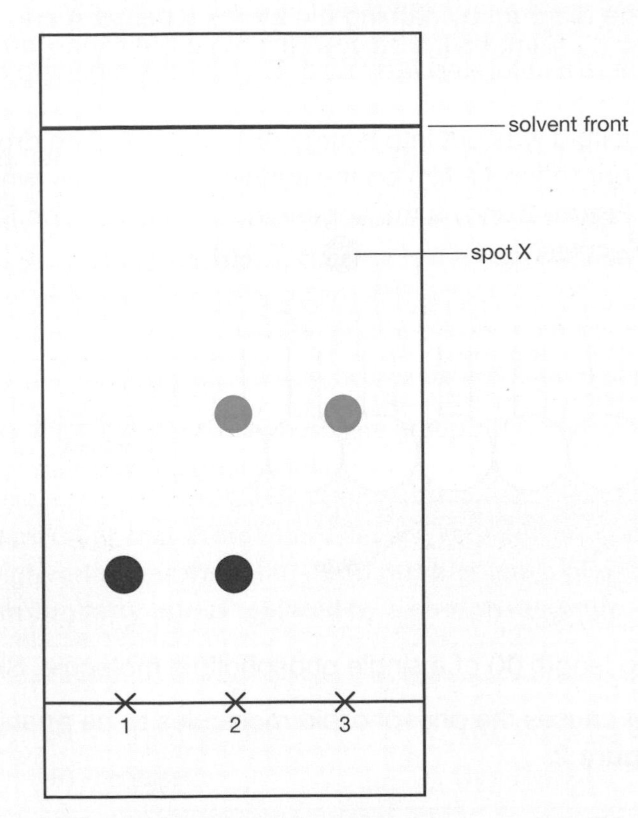

(a) What type of reaction results in the conversion of a dipeptide to amino acids? [1]

(b) Explain why three spots were obtained when the dipeptide was heated with the acid for two minutes. [2]

(c) **(i)** Calculate the R_f value of spot X. Show your working. [2]

(ii) Explain how you could use the R_f values to identify spot X. [1]

[AQA A Specimen]

Answers on pages 13–14 **Answers** on pages 13–14 **Answers** on pages 13–14

(4) (a) Hydrolysis; **(b)** Some of the dipeptide/2 different amino acids;

examiner's tip

Some of the dipeptide was unchanged, hence the spot near the start line. The peptide bond was broken by hydrolysis, into two different amino acids.

(c) (i) 61 mm/76 mm = 0.8 (to 1 d.p.);

examiner's tip

You should measure the distance moved by the spot on the diagram, then measure the distance moved by the solvent front. R_f is distance moved by the spot divided by the solvent front i.e. 8/10 = 0.8

(ii) Check with table of known values or compare with R_f value of known substance.

examiner's tip

R_f value is always the same for the same substance in same solvent under same conditions.

---

(5) (a) (i) 4;

examiner's tip

A triglyceride has 1 glycerol unit + 3 fatty acids.

(ii) Not made of identical units/monomers/made of fatty acids and glycerol;

examiner's tip

Polymers consist of repeated units. Remember that a triglyceride has 1 glycerol unit + 3 fatty acids.

(b) (i) A = oxygen/O; **B** = carbon/C;

examiner's tip

The diagram is an unusual way of representing a saturated fat but you should recognise that the backbone of the molecule is carbon and that the end where A is labelled is a carboxylic acid group, so A is the double bond oxygen position.

(ii) No double bonds/each carbon is attached to 2 hydrogen atoms/every carbon is joined to 4 other atoms;

examiner's tip

The best answer to remember is the lack of C=C double bonds. Next time it may be unsaturated fats which are being tested. They have the C=C double bonds.

(c) (i) Correct answer of 0.0000025/2.5×10^{-6} mm;; (full 2 marks)
Incorrect answer showing volume divided by surface area;

examiner's tip

Never be phased by calculations in biology; they are invariably straightforward. Here the volume of the drop is divided by the surface area of the phospholipid, i.e. 1 mm^3 divided by 400 000 mm^2. It is assumed that the units are in mm. The wrong units would be penalised.

(ii) Head is hydrophilic/attracted to water/polar;
Tail is hydrophobic/avoids water/shuns water/non-polar;

examiner's tip

Here it is important to use the correct terms. Referring to 'loving and hating' water in the correct contexts would score just one mark. Technical words are important, so learn them!

Questions with model answers

C grade candidate – mark scored 4/6

Examiner's Commentary

For help see Revise AS Study Guide pages 33 and 35

1 The diagram below shows the structure of a liver cell as seen using an electron microscope.

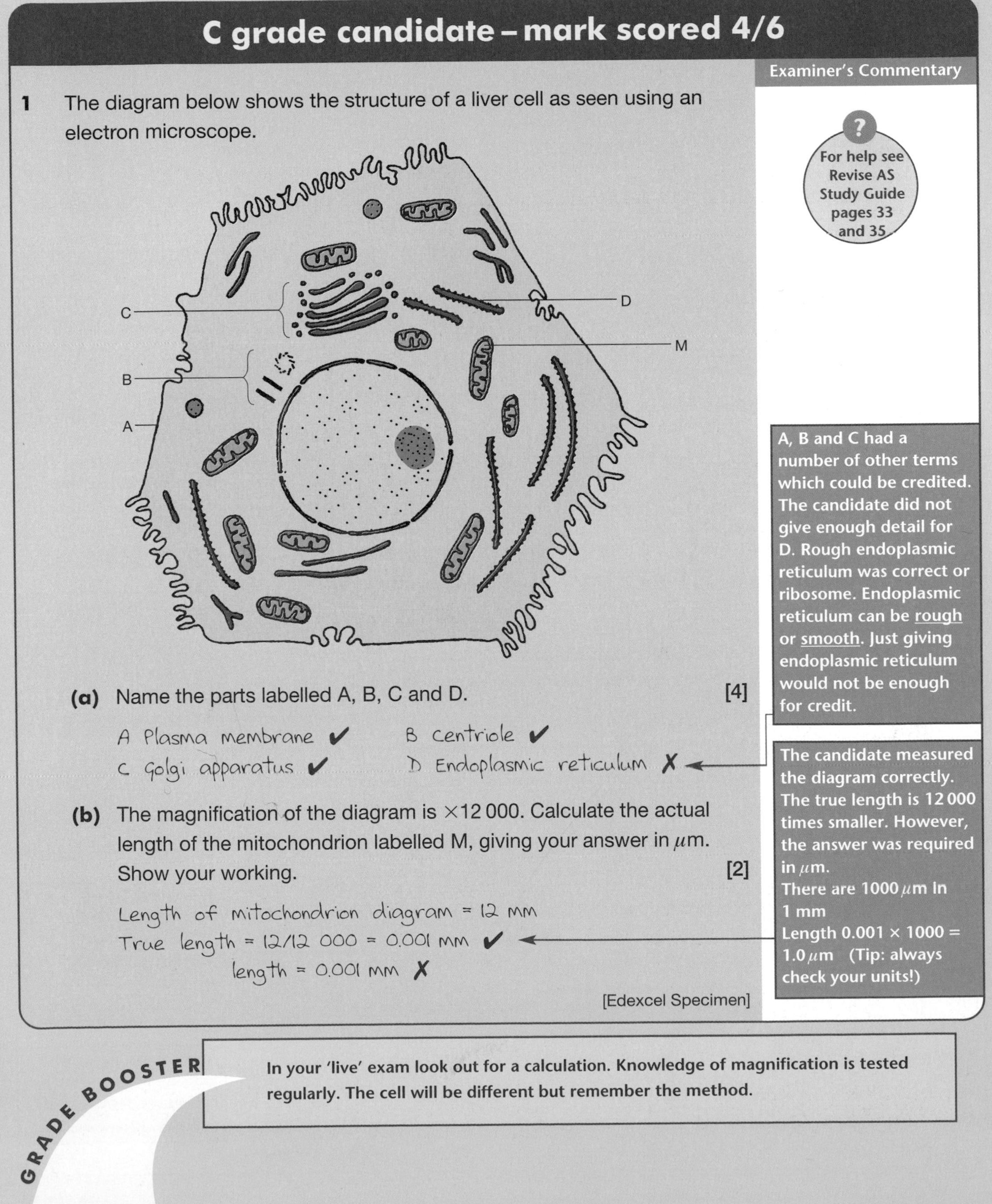

(a) Name the parts labelled A, B, C and D. [4]

A Plasma membrane ✔ B centriole ✔
C Golgi apparatus ✔ D Endoplasmic reticulum ✗

A, B and C had a number of other terms which could be credited. The candidate did not give enough detail for D. Rough endoplasmic reticulum was correct or ribosome. Endoplasmic reticulum can be rough or smooth. Just giving endoplasmic reticulum would not be enough for credit.

(b) The magnification of the diagram is ×12 000. Calculate the actual length of the mitochondrion labelled M, giving your answer in μm. Show your working. [2]

Length of mitochondrion diagram = 12 mm
True length = 12/12 000 = 0.001 mm ✔
length = 0.001 mm ✗

The candidate measured the diagram correctly. The true length is 12 000 times smaller. However, the answer was required in μm.
There are 1000 μm in 1 mm
Length 0.001 × 1000 = 1.0 μm (Tip: always check your units!)

[Edexcel Specimen]

GRADE BOOSTER

In your 'live' exam look out for a calculation. Knowledge of magnification is tested regularly. The cell will be different but remember the method.

A grade candidate – mark scored 12/12

2 The photograph is an electronmicrograph of a mammalian pancreas cell (×15 000).

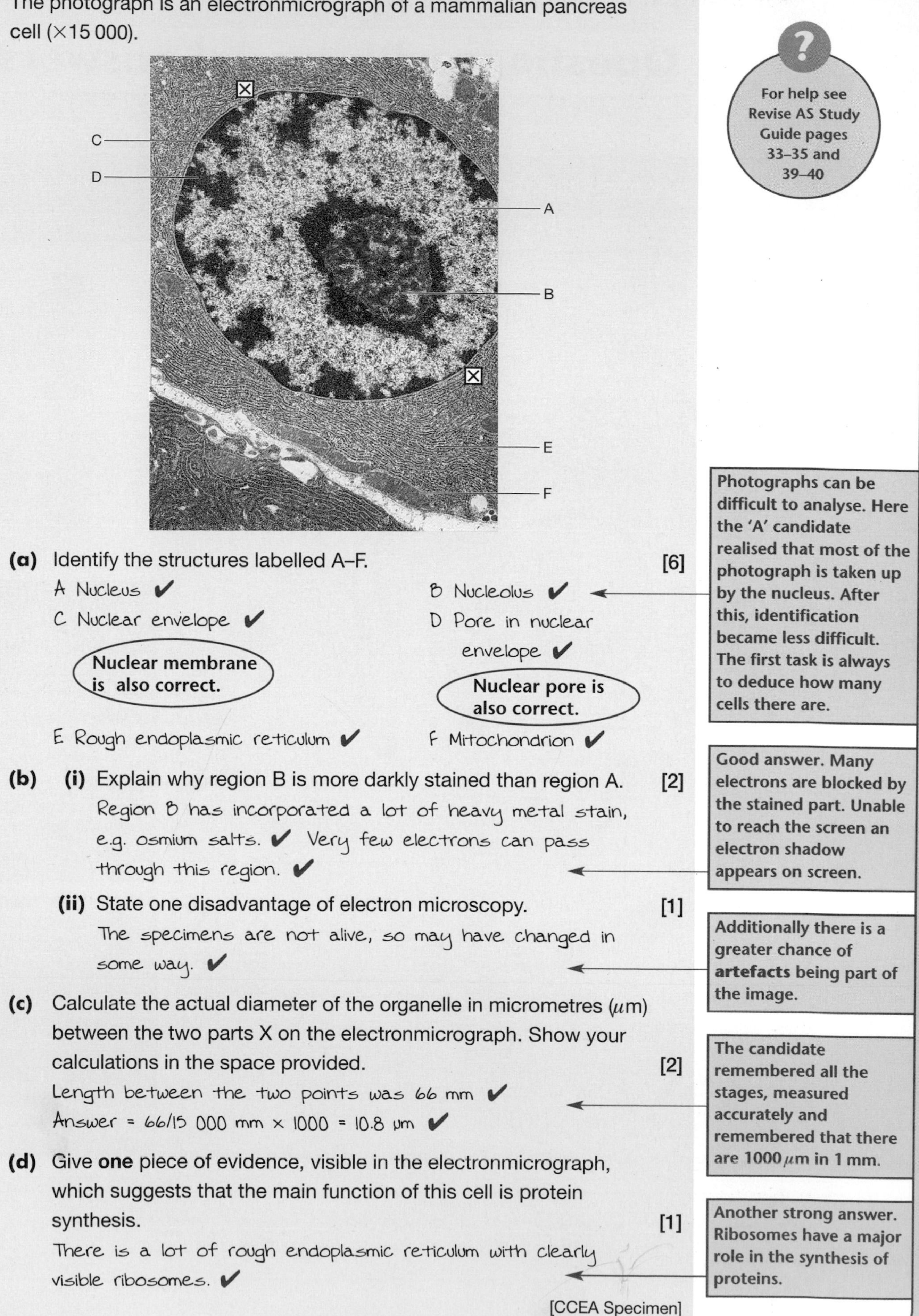

(a) Identify the structures labelled A–F. [6]

A Nucleus ✔
B Nucleolus ✔
C Nuclear envelope ✔
D Pore in nuclear envelope ✔

Nuclear membrane is also correct.

Nuclear pore is also correct.

E Rough endoplasmic reticulum ✔
F Mitochondrion ✔

(b) **(i)** Explain why region B is more darkly stained than region A. [2]

Region B has incorporated a lot of heavy metal stain, e.g. osmium salts. ✔ Very few electrons can pass through this region. ✔

(ii) State one disadvantage of electron microscopy. [1]

The specimens are not alive, so may have changed in some way. ✔

(c) Calculate the actual diameter of the organelle in micrometres (μm) between the two parts X on the electronmicrograph. Show your calculations in the space provided. [2]

Length between the two points was 66 mm ✔
Answer = 66/15 000 mm × 1000 = 10.8 µm ✔

(d) Give **one** piece of evidence, visible in the electronmicrograph, which suggests that the main function of this cell is protein synthesis. [1]

There is a lot of rough endoplasmic reticulum with clearly visible ribosomes. ✔

[CCEA Specimen]

Examiner's Commentary

? For help see Revise AS Study Guide pages 33–35 and 39–40

Photographs can be difficult to analyse. Here the 'A' candidate realised that most of the photograph is taken up by the nucleus. After this, identification became less difficult. The first task is always to deduce how many cells there are.

Good answer. Many electrons are blocked by the stained part. Unable to reach the screen an electron shadow appears on screen.

Additionally there is a greater chance of artefacts being part of the image.

The candidate remembered all the stages, measured accurately and remembered that there are 1000 μm in 1 mm.

Another strong answer. Ribosomes have a major role in the synthesis of proteins.

Exam practice questions

1 Some cells were broken up and the organelles they contained were separated by ultracentrifugation. The drawing shows three types of organelle which were obtained.

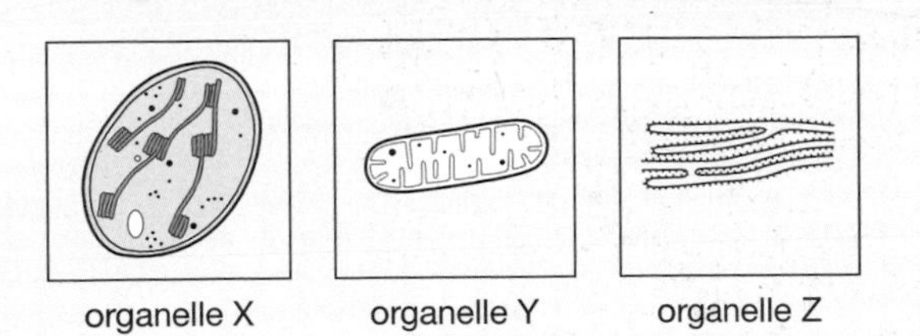

organelle X organelle Y organelle Z

(a) The cells were all the same type. Which of the cells A to D listed below might they have been?

A bacterial cells
B red blood cells
C cells from a plant leaf
D epithelial cells from the lung **[1]**

(b) Explain why only organelle X appeared in the sediment when the broken up cells were centrifuged at the lowest speed. **[1]**

(c) Give the function of:

(i) organelle Y
(ii) organelle Z. **[2]**

(d) Explain why an electron microscope would be required to see the detailed structure of organelle Y. **[2]**

[AQA A Specimen]

---

2 **(a)** The following table lists some of the features of cells. Complete the table by ticking (✔) in the appropriate column(s) if the feature listed is found in eukaryotes, prokaryotes or both cells. **[8]**

Feature	Eukaryotes	Prokaryotes
Usually less than 10 μm in size		
Mitochondria present		
Respiratory enzymes present		
Ribosomes present		
DNA usually a continuous loop		
Presence of nuclear membrane		

(b) Name one way in which the cell wall of most prokaryotic cells differs from that of a plant cell. **[1]**

(c) **(i)** What class of chemical usually forms the outer coat of a virus particle? **[1]**

(ii) What is the coat called? **[1]**

[WJEC Specimen]

Answers on pages 20–21 **Answers** on pages 20–21 **Answers** on pages 20–21

The most widely accepted model of biological molecules is known as the fluid-mosaic model.

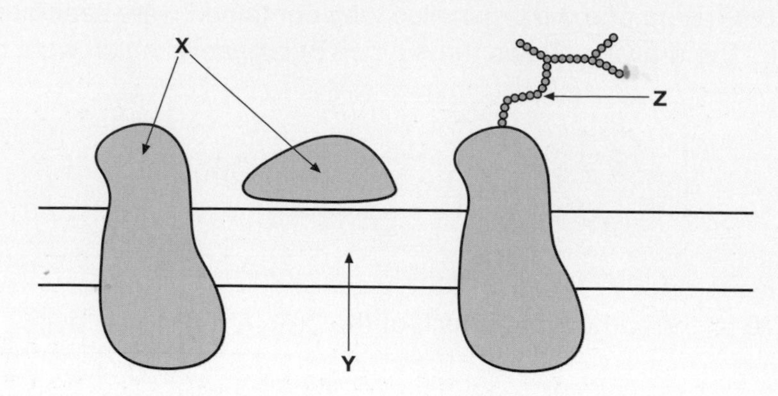

(a) Name the class of biologically important chemicals from which each of the parts labelled **X**, **Y** and **Z** are made. **[3]**

(b) Make a labelled diagram to show the arrangement of the molecules in part **Y**. **[3]**

(c) The model is described as fluid because the component molecules are free to move about. Evidence for this includes experiments like the one shown below, in which component molecules are labelled with a dye.

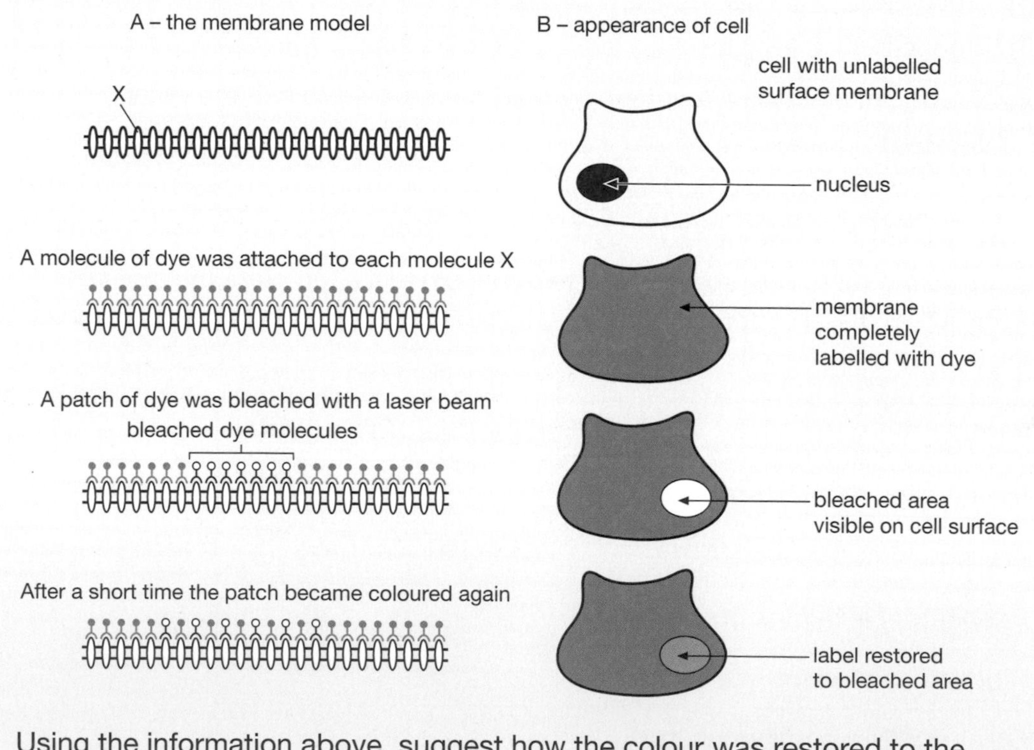

Using the information above, suggest how the colour was restored to the bleached area. **[2]**

(d) Give three functions of the cell surface membrane. **[3]**

[WJEC May 2002]

Cells

Answers on pages 20–21 **Answers** on pages 20–21 **Answers** on pages 20–21

4 The table below refers to three organelles commonly found in eukaryotic cells. Complete the table by writing the name of the organelle, its description or **one** function, as appropriate, in each of the five boxes provided.

Name of the organelle	Description	One function
Golgi apparatus		
	Cylindrical organelles made up of microtubules	Involved in spindle organisation during cell division in animal cells
	Rod-shaped structures with a double membrane, the inner one folded to form cristae	

[5]

[Edexcel June 2001]

Answers

(1) (a) C;

(b) Chloroplasts are the most dense of these organelles and sink first.;

examiner's tip

In this part of the question you need to be aware of the stages of ultracentrifugation. At the slowest speed the nuclei are the most dense and are in the lowest sediment. An assumption you need to make here is that the nuclei must have been removed earlier.

(c) (i) Aerobic respiration; **(ii)** Protein synthesis;

examiner's tip

Learn all of the functions of the organelles. The Exam Board credited respiration but could have insisted on aerobic for part (i). Ribosomes on the rough endoplasmic reticulum clearly point towards protein synthesis.

(d) Needs higher resolution/must distinguish between structures very close together; uses a shorter wavelength (electron beam);

---

(2) (a) Mark each line ticked as follows:
Line 1 P;
Line 2 E;
Line 3 E, P;
Line 4 E, P;
Line 5 P;
Line 6 E;

examiner's tip

This type of question requires you to make a number of decisions. Remember key facts about both eukaryotic and prokaryotic organisms. Prokaryotes do not have membrane bound organelles, so they have no true nucleus or mitochondria.

(b) Plant cell wall is cellulose (bacteria always other chemicals);

(c) (i) Protein; **(ii)** Capsid;

examiner's tip

Outer coat of viruses is made of a number of capsomeres, each of which is a protein unit.

---

(3) (a) **X** = Protein;
Y = (Phospho) lipid;
Z = Carbohydrate/polysaccharide/glycocalyx/glycoprotein;

examiner's tip

The diagram was an unusual representation of the plasma membrane. Molecules X were clearly proteins. Molecules at Y were clearly the phospholipid bilayer. Molecule Z was clearly a recognition protein with its carbohydrate side branch.

(b) Double layer/heads out and tails in; 2 correct labels

examiner's tip

The diagram should be shown as a bilayer of phospholipids. The phospholipid heads point to the outside of the cell and the tails inside next to the cytoplasm.

(c) Labelled molecules/labelled X/labelled protein; reference to mixing or equivalent;

examiner's tip

This question included novel material. A modern technique was used to show the fluid mosaic model of a cell membrane. Interpret information logically! Some of the dye attached to phospholipid molecules lost colour where contact was made with the laser. If the colourless patch remained static then no proof would have been evident. However the patch regained some colour so this proves that the phospholipids were on the move.

(d) Taking up nutrients/other requirements; reference to selective permeability; phagocytosis; secreting chemicals; cell recognition; adhesion; receptor sites;

examiner's tip

Any three of the above answers were acceptable! No logic needed here, just pure recall.
Always remember that the plasma membrane keeps certain substances out of the cell but allows others in, i.e. selective permeability. If you cannot remember the technical terms then an explanation of the function would be acceptable.

---

(4)

Name of the organelle	Description	One function
Golgi apparatus	**Stack/group/of (flattened/curved) cisternae/tubules/sacs;**	**Transport of lipids/ storage of lipids/ modification of lipids/ formation of glycoproteins/ modification of proteins/ formation of secretory vesicles/ formation of lysosomes/ transport of carbohydrates;**
Centrioles/ centrosomes;	Cylindrical organelles made up of microtubules	Involved in spindle organisation during cell division in animal cells
Mitochondria;	Rod-shaped structures with a double membrane, the inner one folded to form cristae	**Aerobic respiration/ ATP production/ oxidative phosphorylation/ Kreb's cycle;**

examiner's tip

You should be ready to answer questions on all organelles found in cells. Key information is supplied so that you can fill in the gaps. The Golgi apparatus packages chemicals such as carbohydrates. The key supplied information of an organelle linked to the spindle pointed to centrioles and centrosomes. Note the description of mitochondria for the final mark.

Chapter 3

Enzymes

Questions with model answers

C grade candidate – mark scored 8/11

1 (a) Define the term 'enzyme'. [2]

An enzyme is a molecule which catalyses reactions in living organisms ✔ by lowering the activation energy. ✔

(b) Explain why enzymes are essential for metabolism. [2]

Lowering the activation energy allows reactions to take place at the usual physiological temperature of the organism, such as 37 °C in humans ✔.

(c) The diagram shows the influence of substrate concentration on the rate of an enzyme-controlled reaction working at its **optimum temperature.**

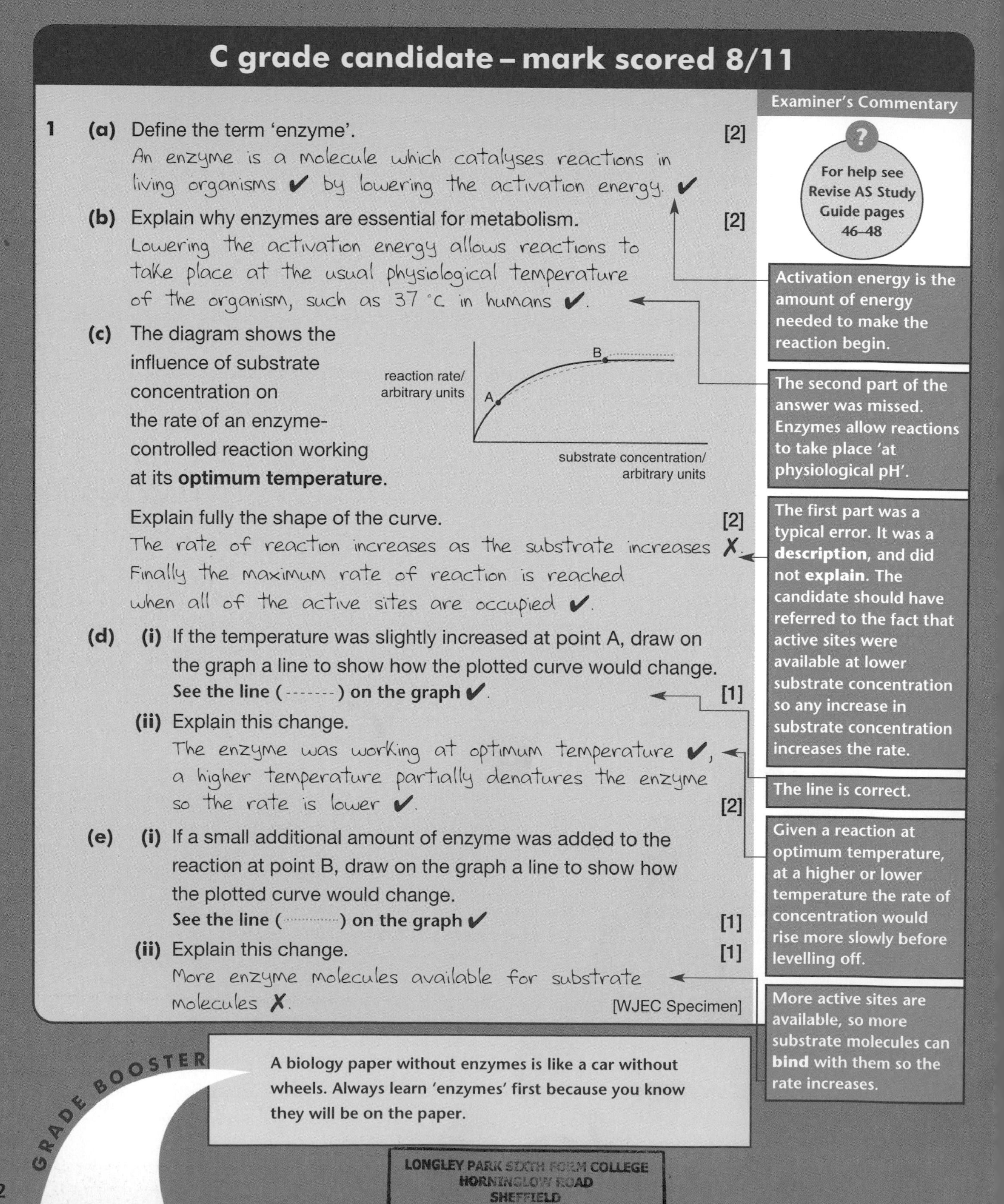

Explain fully the shape of the curve. [2]

The rate of reaction increases as the substrate increases ✗. Finally the maximum rate of reaction is reached when all of the active sites are occupied ✔.

(d) (i) If the temperature was slightly increased at point A, draw on the graph a line to show how the plotted curve would change.

See the line (-------) on the graph ✔. [1]

(ii) Explain this change.

The enzyme was working at optimum temperature ✔, a higher temperature partially denatures the enzyme so the rate is lower ✔. [2]

(e) (i) If a small additional amount of enzyme was added to the reaction at point B, draw on the graph a line to show how the plotted curve would change.

See the line (..............) on the graph ✔ [1]

(ii) Explain this change. [1]

More enzyme molecules available for substrate molecules ✗.

[WJEC Specimen]

Examiner's Commentary

?
For help see Revise AS Study Guide pages 46–48

Activation energy is the amount of energy needed to make the reaction begin.

The second part of the answer was missed. Enzymes allow reactions to take place 'at physiological pH'.

The first part was a typical error. It was a **description**, and did not **explain**. The candidate should have referred to the fact that active sites were available at lower substrate concentration so any increase in substrate concentration increases the rate.

The line is correct.

Given a reaction at optimum temperature, at a higher or lower temperature the rate of concentration would rise more slowly before levelling off.

More active sites are available, so more substrate molecules can **bind** with them so the rate increases.

GRADE BOOSTER

A biology paper without enzymes is like a car without wheels. Always learn 'enzymes' first because you know they will be on the paper.

A grade candidate – mark scored 9/10

Examiner's Commentary

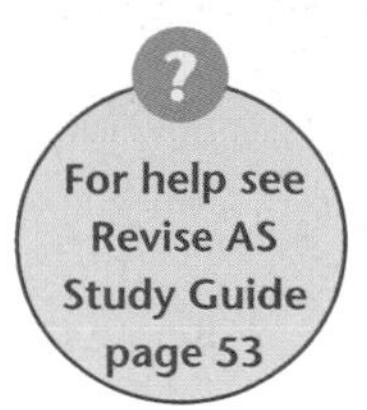

2 (a) The diagrams below represent two alternative commercial production systems involving enzymes. In the batch reactor, a fixed amount of soluble enzyme and substrate are mixed together in a solution. In the continuous-flow column reactor, substrate molecules flow past enzymes which are immobilised on an inert support material.

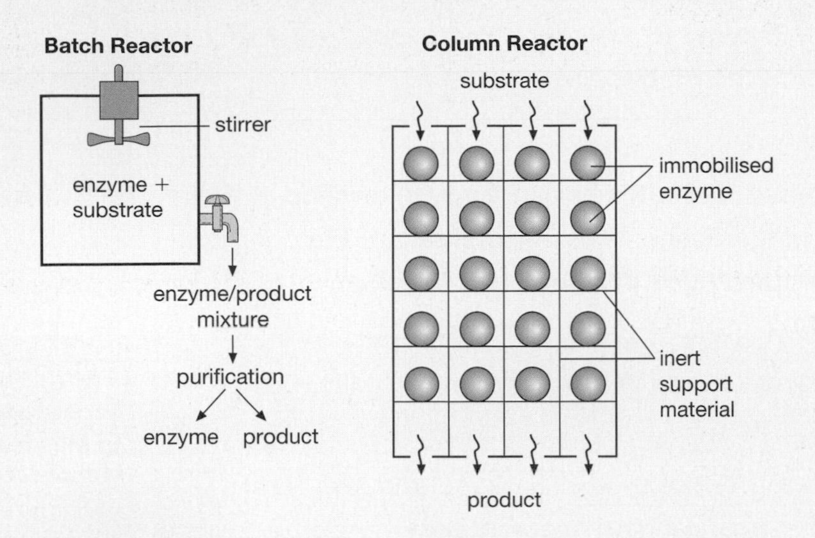

Identify **two** advantages that the immobilised enzyme system has over the dissolved enzyme system. Explain your answer. [4]

Immobilised enzymes adhere to the inert support material, so can be used continuously, ✔ because the enzymes remain, this is more economic. ✔ The enzymes are withheld in the reactor and so do not contaminate the product ✔ so no purification is needed. ✔

Often, immobilised enzymes work just as well as freely moving ones. Here the graph suggests that some of the active sites of some enzyme molecules have been affected.

(b) One disadvantage of enzyme immobilisation is that it may change the catalytic activity of the enzyme as shown in the graph below.

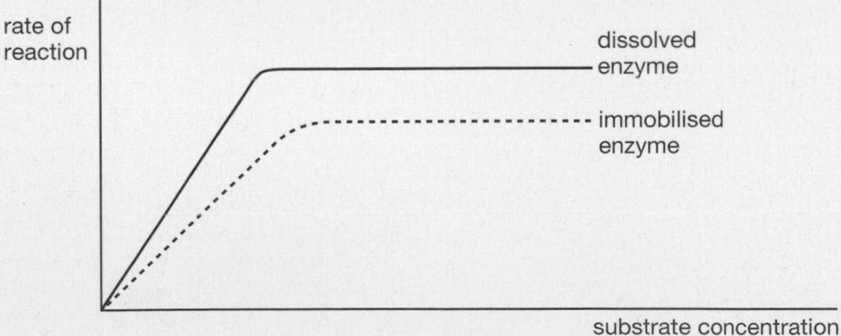

Enzymes

Suggest **two** reasons for the observed change in enzyme activity when immobilised. [4]

The active sites of some enzyme molecules could be covered by an 'adhesive' material ✔, so that the binding of some substrate molecules could be prevented ✔. As the substrate molecules flow past the enzymes there is less of a chance for the substrate molecules to bind with the active sites ✔, so there is a slower rate of enzyme-substrate complex ✔.

(c) Immobilisation is generally accepted as a strategy to develop more stable enzyme preparations. Thus immobilisation often increases the temperature and/or pH range over which an enzyme remains active.

Suggest how immobilisation may improve enzyme stability. Explain your answer. [2]

There is strong bonding between the enzyme and the inert support material, ✔ therefore the enzyme does not change ✗.

[CCEA Specimen]

Examiner's Commentary

Full marks! As the substrate molecules flow along there is a tendency for the rate to be reduced if it is more difficult to contact an active site.

First part correct. Second part not correct. Immobilisation helps withstand the disruption of the tertiary structure of the enzyme by higher temperature or pH change.

Exam practice questions

1 Catalase is an enzyme that breaks down hydrogen peroxide into oxygen and water. The activity of catalase can be measured by soaking small discs of filter paper in a solution containing the enzyme. The discs are immediately submerged in a dilute solution of hydrogen peroxide. The filter paper discs sink at first but float to the surface as oxygen bubbles are produced. The reciprocal of the time taken for the discs to rise to the surface indicates the rate of reaction.

An experiment was carried out to investigate the effect of substrate concentration on the activity of catalase. A filter paper disc was soaked in a solution containing catalase, and then submerged in a buffer solution containing hydrogen peroxide. The time taken for the disc to rise to the surface was recorded. The experiment was repeated using a range of concentrations of hydrogen peroxide.

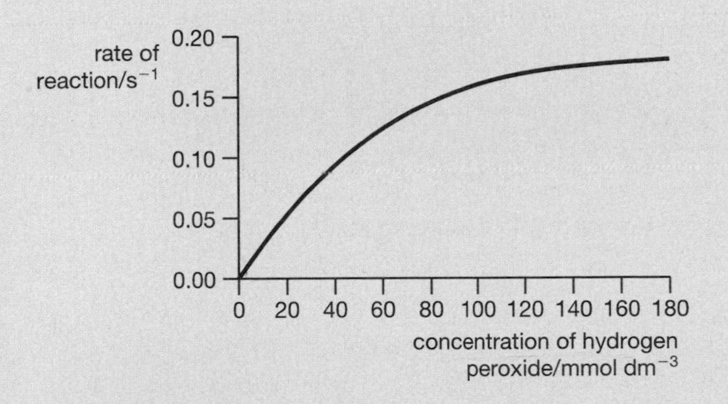

(a) State why a buffer solution was used in this experiment. [1]

(b) Describe the relationship between the rate of reaction and the concentration of hydrogen peroxide. [3]

(c) Explain this relationship between substrate concentration and the rate of reaction. [4]

(d) Describe how a solution containing 160 mmol of hydrogen peroxide per dm^3 would be diluted to prepare a solution containing 80 mmol of hydrogen peroxide per dm^3. [2]

(e) Describe how this experiment could be modified to investigate the effect of temperature on the activity of catalase. [2]

---

2 **(a)** Complete the series of diagrams below to show how a catabolic reaction would take place, in terms of the induced fit theory. Label your drawings. [3]

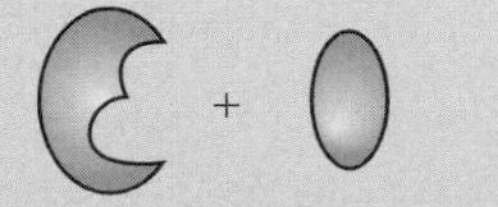

(b) Explain TWO ways that an enzyme catalysed reaction can be prevented by the presence of molecules other than the substrate. [3]

Answers on pages 27–28 **Answers** on pages 27–28 **Answers** on pages 27–28

(a) Triglycerides can be broken down without an enzyme being present. The graph shows the energy changes that take place during this reaction.

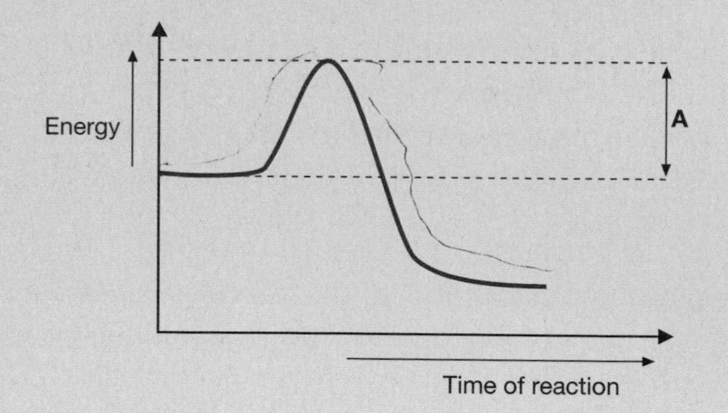

(i) What is represented by **A**? [1]

(ii) Sketch a curve on the graph to show the energy changes that take place when lipase is present. [2]

(iii) A test tube containing a mixture of triglyceride and lipase was incubated at 35 °C. Explain what would happen to the pH in the test tube. [2]

(b) The diagram shows the human digestive system.

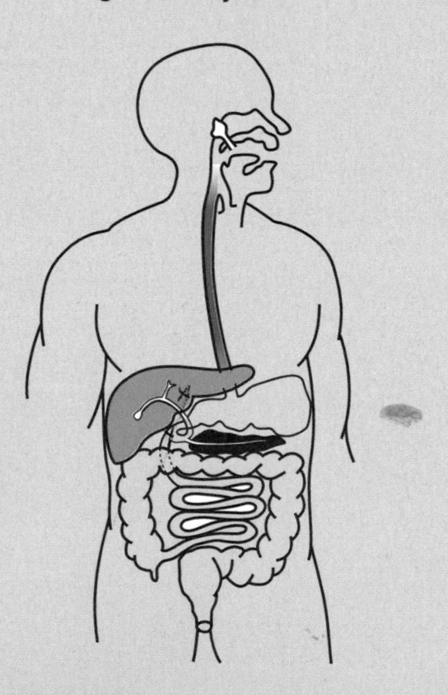

(i) Label with a guideline and the letter X, an organ that produces an endopeptidase. [1]

(ii) Label with a guideline and the letter Y, an organ in which bile is produced. [1]

(iii) Describe one role of bile in the digestion of triglycerides. [2]

[AQA B May 2002]

Enzymes

Answers on pages 27–28 **Answers** on pages 27–28 **Answers** on pages 27–28

Answers

(1) (a) To keep pH constant/enzymes are affected by pH.;

> **examiner's tip**
>
> **Always be prepared to use the term 'buffer'. It keeps the pH constant even though more H^+ ions would increase acidity or more OH^- ions would increase alkalinity.**

(b) Rate increases as the concentration of substrate increases;
There is a constant increase (approximately) between 20–80 mmol dm^{-3};
Rate levels off after this;

(c) (In terms of kinetic theory) number of collisions increase with the increase of substrate concentration; as a result the rate increases; until all of the active sites are in use; rate is then constant or maximum;

> **examiner's tip**
>
> **Collision theory is about the random impacts between the active site of the enzymes and substrate molecules. As the number of substrate molecules increases the chance of suitable collisions increases.**

(d) Mix equal volumes; of hydrogen peroxide solution and distilled water or deionised water or buffer solution.;

(e) Suggest a range of temperatures (at least three); use of same substrate concentration; use the same volume of hydrogen concentration (standardise pH); allow the substrate to equilibrate before adding the filter paper disc; discs must be uniform in size; use of same enzyme concentration; repeat at each temperature; plot a graph of the rate of reaction against temperature.;

> **examiner's tip**
>
> **This tests a key skill, your ability to design or modify an investigation. Note that there is usually a mark for replication. Here a range of at least three temperatures must be suggested, to earn a mark. Then another mark is available for repeating each chosen temperature. Remember this for your live examination!**

---

(2) (a) Draw the enzyme with active site changed to suit substrate shape; label the enzyme substrate complex; two or more products formed.;

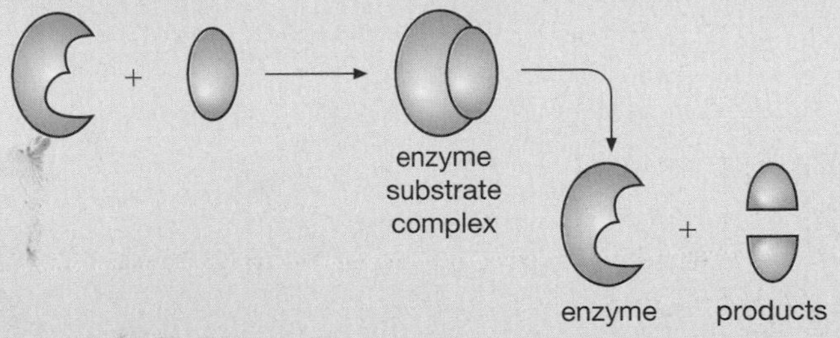

(b) Competitive inhibitor, molecules fit into active site preventing substrate entry.;
Non-competitive inhibitor, molecules fit into alternative site, change shape of active site preventing substrate entry.;
H^+ or OH^- ions or other correctly named substance, which can denature enzyme changing active site.;

> **examiner's tip**
>
> **Always remember that an inhibitor can prevent the reaction or reduce the rate of reaction.**
> **A non-competitive inhibitor binds to a different part of the enzyme than the active site but it still changes the shape of the active site.**

Enzymes

(3) (a) (i) Activation energy;

examiner's tip

Activation energy input is needed to begin the reaction. Once this threshold is reached or exceeded then the reaction takes place.

(ii) Activation energy is lower, but within a peak;
Curve must start and finish at original energy levels;
(Higher curve disqualifies the mark)

examiner's tip

This is a classic question (regularly asked!). An enzyme catalysed reaction would have a lower peak on the graph. The start and end of your sketch must start at the same point as the supplied graph does and end at the same level as the end of the supplied graph.

(iii) pH drops or lowers/it becomes acidic;
Fatty acids are formed (when lipase breaks down lipids);

examiner's tip

Triglyceride is a lipid so would break down to fatty acids and glycerol. It is the fatty acids formation in the tube which drop the pH.

(b) (i) An X placed on the stomach or pancreas;

(ii) A Y placed on the liver; (reject gall bladder)

examiner's tip

There was danger here in losing the mark. The liver produces the bile, and the gall bladder merely stores it. Avoid the Y touching the 'sac' part in the liver (attached to bile duct) as it represents the gall bladder!

(iii) Emulsifies/produces small droplets (of triglycerides);
increases surface area/large surface area available (for digestion);
OR
Neutralises (stomach) acid;
optimum pH for lipase/enzymes; (incorrect enzymes disqualify the mark)

examiner's tip

Emulsification is the break up of lipids into tiny globules of the same substance. If you imply breakdown into another substance then you lose the mark.

Chapter 4 Exchange and transport

Questions with model answers

C grade candidate – mark scored 5/7

1 The diagram below shows a human heart at a specific stage in the cardiac cycle.

(a) Name the parts labelled A and B. [2]

A Semi-lunar valve ✔ B Mitral valve ✔

(b) Name the stage of the cardiac cycle shown in the diagram and give TWO reasons for your choice. [3]

Name of stage: systole ✗

Reason 1: the mitral valve is open ✔.

Reason 2: the semi-lunar valve is closed ✔.

(c) Give ONE function of each of the parts X and Y. [2]

X prevents the valve being inverted, back into the atrium ✔.

Y is an anchor point for the tendonous cords ✗.

[Edexcel Specimen]

Examiner's Commentary

For help see Revise AS Study Guide pages 66–68

Both answers are correct. A is also called a pocket valve but was not in the mark scheme of this Exam Board. B is also called the left atrioventricular valve or the bicuspid valve.

The response is not accurate enough. **Atrial** systole would have been credited because the atria are contracting.

Both of the reasons are correct and can be deduced from the diagram.

The function of X is correct. The tendonous cords prevent the valve from collapsing back into the atrium. The response to the function of Y gains no credit. It adjusts the tension in the valve and actually contracts to cause this.

GRADE BOOSTER

Note the **detail** needed in the answers. This is the key. You can only improve with regular revision.

A grade candidate – mark scored 14/15

(Quality of written communication will be assessed in this question)

2 (a) Mesophyll cells in a plant use carbon dioxide for photosynthesis. Describe how carbon dioxide from the atmosphere reached these mesophyll cells. [2]

Carbon dioxide enters the leaf by diffusion ✔, through the stomata and the air spaces ✔.

(b) The diagram below shows the way in which water flows over the gills of a fish. The graph shows the changes in pressure in the mouth cavity and in the opercular cavity during a ventilation cycle.

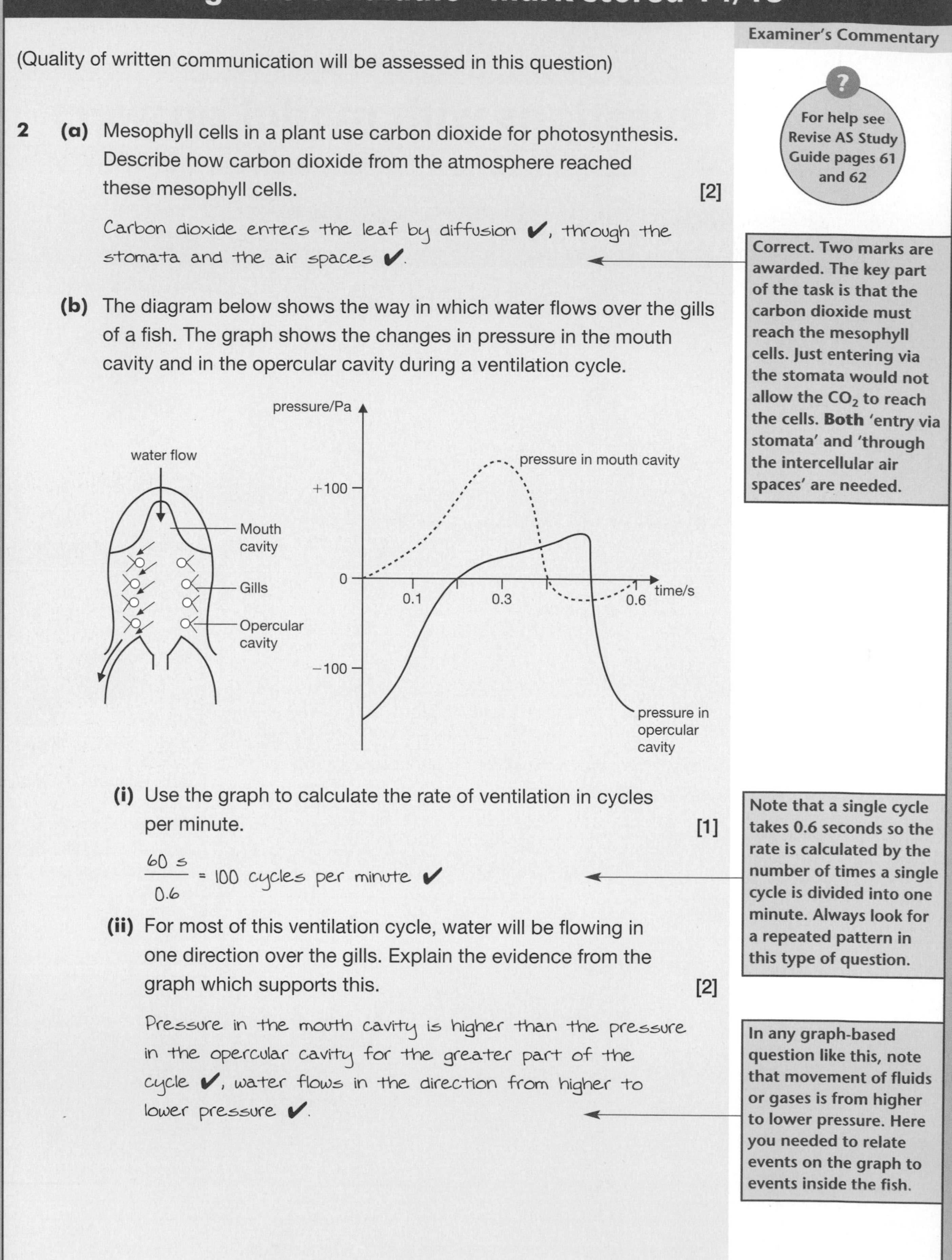

(i) Use the graph to calculate the rate of ventilation in cycles per minute. [1]

$\frac{60\ s}{0.6}$ = 100 cycles per minute ✔

(ii) For most of this ventilation cycle, water will be flowing in one direction over the gills. Explain the evidence from the graph which supports this. [2]

Pressure in the mouth cavity is higher than the pressure in the opercular cavity for the greater part of the cycle ✔, water flows in the direction from higher to lower pressure ✔.

Examiner's Commentary

?

For help see Revise AS Study Guide pages 61 and 62

Correct. Two marks are awarded. The key part of the task is that the carbon dioxide must reach the mesophyll cells. Just entering via the stomata would not allow the CO_2 to reach the cells. Both 'entry via stomata' and 'through the intercellular air spaces' are needed.

Note that a single cycle takes 0.6 seconds so the rate is calculated by the number of times a single cycle is divided into one minute. Always look for a repeated pattern in this type of question.

In any graph-based question like this, note that movement of fluids or gases is from higher to lower pressure. Here you needed to relate events on the graph to events inside the fish.

Examiner's Commentary

(iii) Explain how the fish increases pressure in the buccal cavity. [2]

The mouth closes ✔.

Muscles at the base of the mouth contract to allow the base to move upwards ✔.

> Remember to work out the answer if you do not know it directly. Pressure could only build up if the mouth is shut.

(c) Explain how the gills of a fish are adapted to form a specialised exchange surface. [8]

The gill filaments are very thin ✔, and there are many of them ✔.

Each filament is lined with flattened epithelial cells ✔, which gives a short diffusion path ✔.

Water moves in the opposite direction to blood ✔, which maximises the diffusion gradient along each filament ✔.

The ventilation movements of the buccal cavity and opening and closing of the operculum continually allows water to cross the gills ✔.

> The candidate answered well but just fell short of the 8 marks available. The candidate was aware of the replacement of water but failed to describe the replacement of blood to the gills by the circulatory system. Deoxygenated blood returning to the gills allows a maximum rate of gaseous exchange.

[AQA B Specimen]

Exam practice questions

1 **(a)** **(i)** Define the term 'water potential'. **[1]**

(ii) What is the water potential of pure water? **[1]**

(b) The graph shows the change in cell volume and water potential values of a flaccid plant cell placed in distilled water.

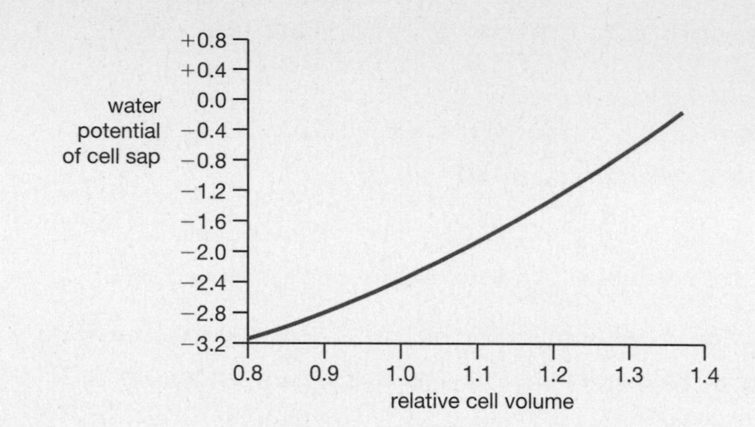

Using the graph,

(i) Describe the change in water potential of the cell. **[1]**

(ii) Explain your answer. **[1]**

(c) The **solute** potential of the cell is never zero. Explain. **[1]**

[WJEC Specimen]

---

2 The diagram below shows a section through the heart of a mammal.

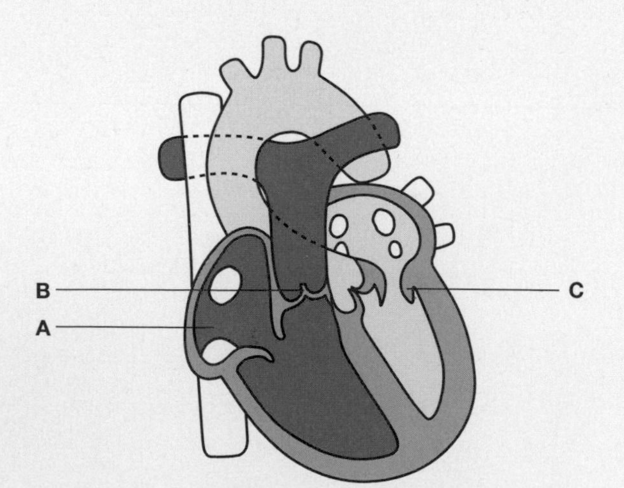

(a) Name the parts labelled **A**, **B** and **C**. **[3]**

(b) Each time the heart beats, the atria contract first and then the ventricles contract. Explain how this sequence of events is coordinated. **[4]**

[Edexcel June 2001]

Answers on pages 35–36 **Answers** on pages 35–36 **Answers** on pages 35–36

3 The mean concentration of sugar in the phloem sap of the stems of cotton plants was measured at four different times of the day and the results are plotted in the graph **(C)** below.

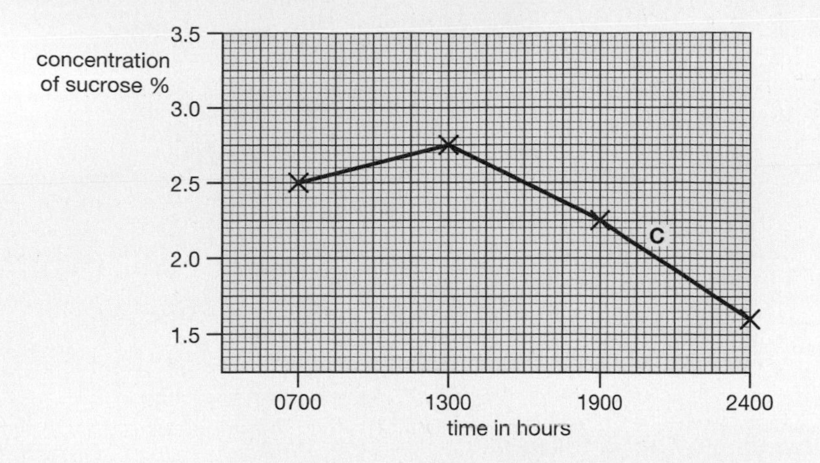

At the start of the experiment, a second batch of plants had a ring of tissue removed from the stem, destroying the phloem at the point where the ring was taken. The table shows the sugar concentration above the ring **(A)** and the concentrations below the ring **(B)**, at the same time intervals.

	Time/hours			
	0700	**1300**	**1900**	**2400**
A (above)	2.5	3.5	3.1	2.7
B (below)	2.5	1.9	1.6	1.5

(a) Name the conducting vessels in the phloem that were destroyed by the ringing process. [1]

(b) Plot the values for **(A)** and **(B)** on the graph above. [2]

(c) For the period 0700 hours to 1300 hours, explain **fully**:

(i) the difference between **(A)** and **(C)** [2]

(ii) the difference between **(B)** and **(C)**. [2]

(d) What is the reason for the trend shown by the plots between 1900 hours and 2400 hours? [1]

[WJEC Specimen]

4 The diagram below shows the formation of tissue fluid from part of a capillary.

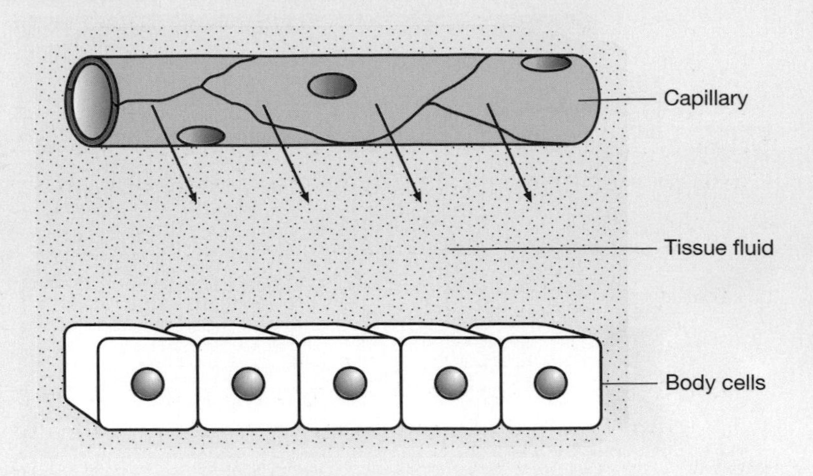

(a) Describe how tissue fluid is formed. [2]

(b) The table below shows the concentration of some solutes in blood plasma and tissue fluid.

	Concentration in mmol dm^{-3}	Concentration in mmol dm^{-3}
Potassium ions	4.0	4.0
Sulphate ions	0.5	0.5
Protein	2.0	Less than 0.1

(i) Compare the concentration of these solutes in blood plasma and tissue fluid. [2]

(ii) Suggest explanations for the difference in the concentration of these solutes in blood plasma and tissue fluid. [3]

[Edexcel 2B June 2001]

Answers on pages 35–36 **Answers** on pages 35–36 **Answers** on pages 35–36

Answers

(1) (a) (i) Water potential is the tendency of water to leave the cell.;

(ii) Zero;

examiner's tip

There must be a solute, such as mineral salts, in the water if water potential is to have a value. This being so, it would be negative and the usual units would be pascals.

(b) (i) As the water potential of the cell becomes less negative the cell volume increases.;

(ii) Water enters the sap vacuole by osmosis, because the cell sap is hypertonic to the distilled water, as this happens the water potential changes.;

examiner's tip

Always remember that water moves from areas of a less negative water potential to a more negative water potential. As water enters a cell the water potential value becomes less negative.

(c) Cell sap always contains some mineral salts.;

---

(2) (a) **A** = right atrium; (accept right auricle);
B = (pulmonary) semilunar valve;
C = mitral/bicuspid/(left) atrioventricular valve;

examiner's tip

Learn the heart structures; you know that names will be tested! Remember that the view you are given is invariably the same way around, ie. your left is the right of the heart.

(b) 1 Reference to pacemaker/sino-atrial node/SAN;
2 (Wave of) excitation/in (walls of) atria;
3 causes contraction (of muscle) in atria (walls);
4 Delay at, atrioventricular node/AVN;
5 Conducted to ventricles via bundles of His/Purkyne fibres; (Any 4)

examiner's tip

This is a classic question! Learn the sequence from SAN through to contraction of the ventricles. Not appropriate here, but usually included is stimulation/inhibition via sympathetic nerve and vagus nerve respectively.

(3) (a) Sieve tubes;

(b) One mark is taken off for each plotting error.

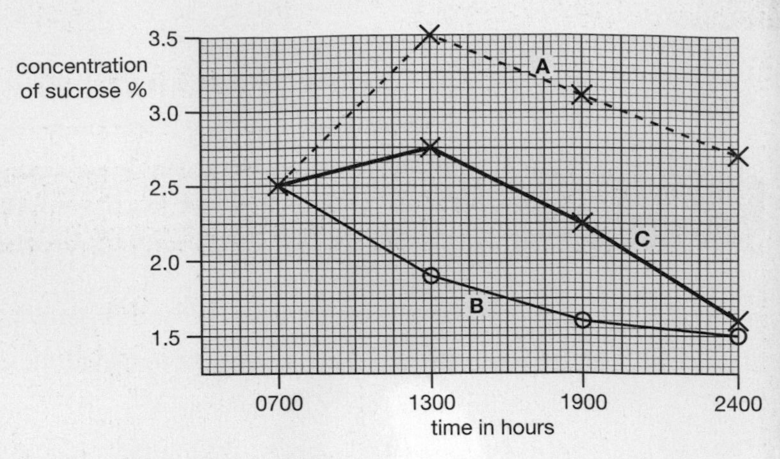

(c) (i) The sugar accumulates above the cut/ringing does not allow the sugar to pass down; (the sugar is not being used quickly so production of sugar, by photosynthesis, exceeds the amount being used in respiration).;

(ii) The supply of sugar has been cut off by the ringing technique; and sugar is being used in respiration so sugar concentration falls.;

examiner's tip

Ringing destroys the continuity of the phloem elements. The leaves above the damage continue to photosynthesise and so sugar production accumulates over the ringed part. Below, the roots have a big energy requirement, so sugar concentration drops quickly.

(d) Photosynthesis gradually stops, so the sugar concentration falls as the sugar is utilised in respiration.;

---

(4) (a) Blood pressure forces water out of capillary;
the pressure is hydrostatic;
capillary walls are permeable to water;

examiner's tip

The arteriole end of capillary bed favours the movement of solutes such as oxygen to the tissue fluid and into capillaries at the venule end. Note the detail that is required. You must avoid vague answers!

(b) (i) Sulphate ions and potassium ions are each equal in concentration in plasma and tissue fluid;
Many more protein ions in plasma than tissue fluid;

examiner's tip

Where the sulphate ions and potassium ions are equal in concentration they diffuse in both directions equally, showing zero net movement.

(ii) Protein molecules too big to pass through the capillary wall/capillary wall impermeable to protein molecules;
protein needs to remain in plasma;
plasma proteins reduce water potential of the blood;
to reabsorb water from tissue fluid; (Any 3)

examiner's tip

The key fact is the size of the protein molecules being too big to escape through the capillary wall. The production of tissue fluid is not usually explained well. Give extra revision time.

Chapter 5

The genetic code

Questions with model answers

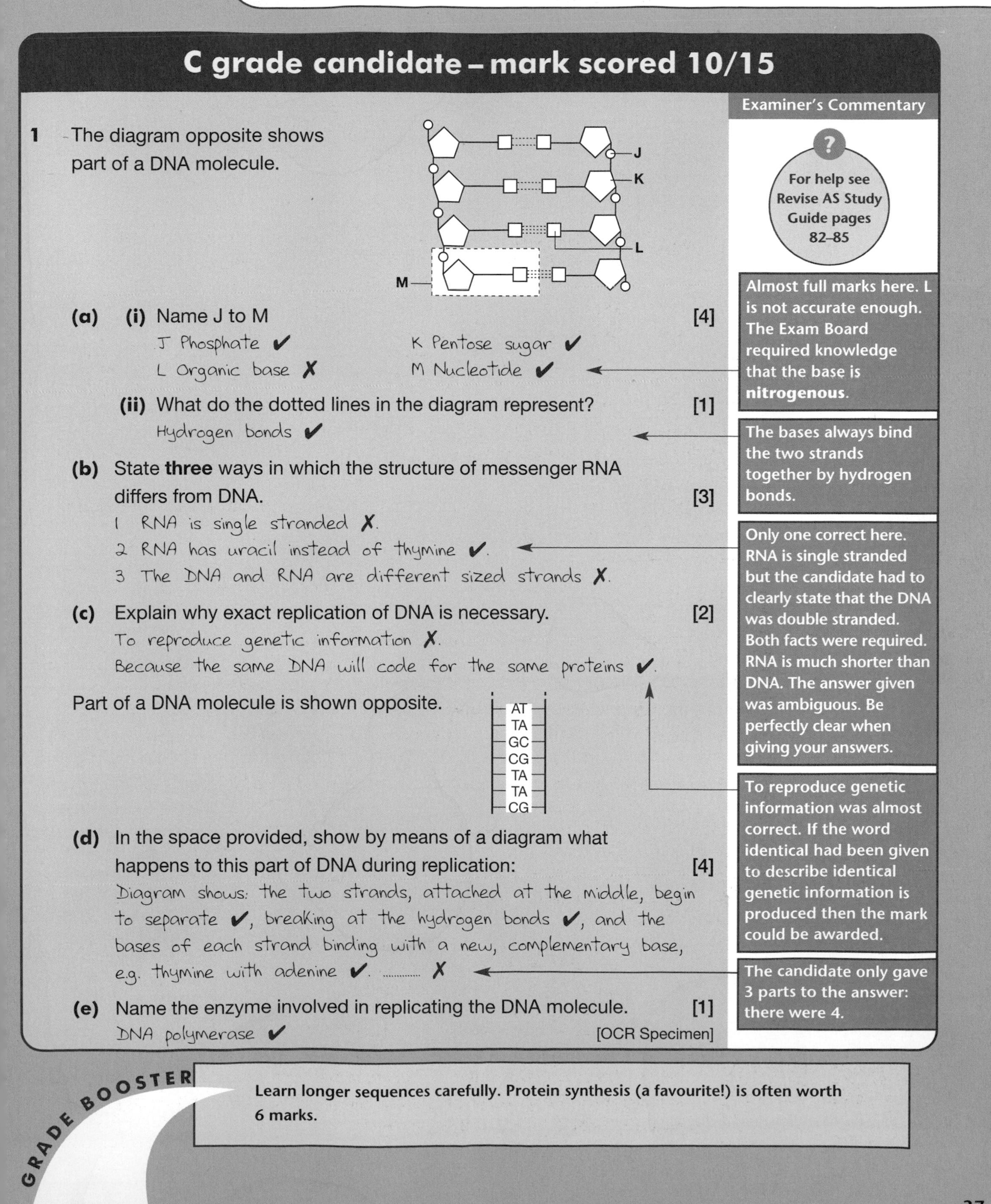

C grade candidate – mark scored 10/15

1 The diagram opposite shows part of a DNA molecule.

(a) (i) Name J to M [4]

J Phosphate ✔ *K Pentose sugar ✔*

L Organic base ✗ *M Nucleotide ✔*

(ii) What do the dotted lines in the diagram represent? [1]

Hydrogen bonds ✔

(b) State **three** ways in which the structure of messenger RNA differs from DNA. [3]

1 RNA is single stranded ✗.

2 RNA has uracil instead of thymine ✔.

3 The DNA and RNA are different sized strands ✗.

(c) Explain why exact replication of DNA is necessary. [2]

To reproduce genetic information ✗.

Because the same DNA will code for the same proteins ✔.

Part of a DNA molecule is shown opposite.

(d) In the space provided, show by means of a diagram what happens to this part of DNA during replication: [4]

Diagram shows: the two strands, attached at the middle, begin to separate ✔, breaking at the hydrogen bonds ✔, and the bases of each strand binding with a new, complementary base, e.g. thymine with adenine ✔. .... ✗

(e) Name the enzyme involved in replicating the DNA molecule. [1]

DNA polymerase ✔

[OCR Specimen]

Examiner's Commentary

?

For help see Revise AS Study Guide pages 82–85

Almost full marks here. L is not accurate enough. The Exam Board required knowledge that the base is **nitrogenous**.

The bases always bind the two strands together by hydrogen bonds.

Only one correct here. RNA is single stranded but the candidate had to clearly state that the DNA was double stranded. Both facts were required. RNA is much shorter than DNA. The answer given was ambiguous. Be perfectly clear when giving your answers.

To reproduce genetic information was almost correct. If the word identical had been given to describe identical genetic information is produced then the mark could be awarded.

The candidate only gave 3 parts to the answer: there were 4.

GRADE BOOSTER

Learn longer sequences carefully. Protein synthesis (a favourite!) is often worth 6 marks.

A grade candidate – mark scored 8/8

Examiner's Commentary

For help see Revise AS Study Guide pages 87–114

2 Biologists in Australia are using genetic engineering to produce a gene to insert into oranges. They have combined two pieces of DNA to produce a gene which they have called SDLS-2.

The diagram shows part of this gene.

DNA sequence that switches on a gene used in seed formation	DNA sequence that kills plant cells

They intend to introduce this gene into orange trees.

(a) Describe how each of the following might be useful in this process.

(i) Ligase enzymes [1]

The ligase enzymes join the DNA pieces together ✔.

> In genetic engineering there are a limited number of enzymes to remember. Note that ligase enzymes attach compatible ends of DNA together.

(ii) A vector [2]

The new gene is inserted into a vector first ✔, which is then taken up by a plant cell ✔.

(b) **(i)** Explain how this gene could lead to the production of seedless fruit. [2]

The gene is switched on as seeds are formed ✔, so that the seed is finally killed during formation ✔.

> Full marks. Additionally the fact that the fruit cells are still able to develop, would have been credited.

(ii) Describe the possible dangers which might result from growing orange trees containing the SDLS-2 gene. [3]

The plants no longer have seeds, but they are needed for reproduction ✔.

If the genes were able to become incorporated into other species, then they would become sterile ✔.

> This is the most difficult part of the question. Analyse information like this, and give good ideas associated with the problem. It was logical to suggest that the gene may reach another species and the consequences of this, e.g. the onset of sterility.

If the gene which kills cells was switched on in other parts of a plant then it is likely that the complete plant would be killed ✔.

[AQA Specimen]

Exam practice questions

1 **(a)** The drawing shows part of a DNA molecule.

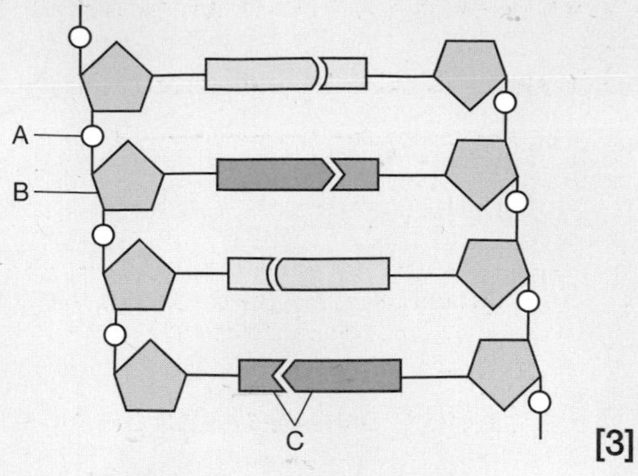

Name the parts labelled. [3]

(b) The text below shows the sequence of bases in a short length of mRNA.

A U G G C C U C G A U A A C G G C C A C C A U G

(i) What is the maximum number of amino acids in the polypeptide for which this piece of mRNA could code? [1]

(ii) How many different types of tRNA molecule would be used to produce a polypeptide from this piece of mRNA? [1]

(iii) Give the DNA sequence which would be complementary to the first five bases in this piece of mRNA. [1]

(c) Name the process by which mRNA is formed in the nucleus. [1]

(d) Give one way in which the structure of a molecule of tRNA differs from the structure of a molecule of mRNA. [1]

[AQA B Specimen]

2 The diagrams show six stages of mitosis, labelled A to F, in a plant root tip, as seen under high power of a light microscope.

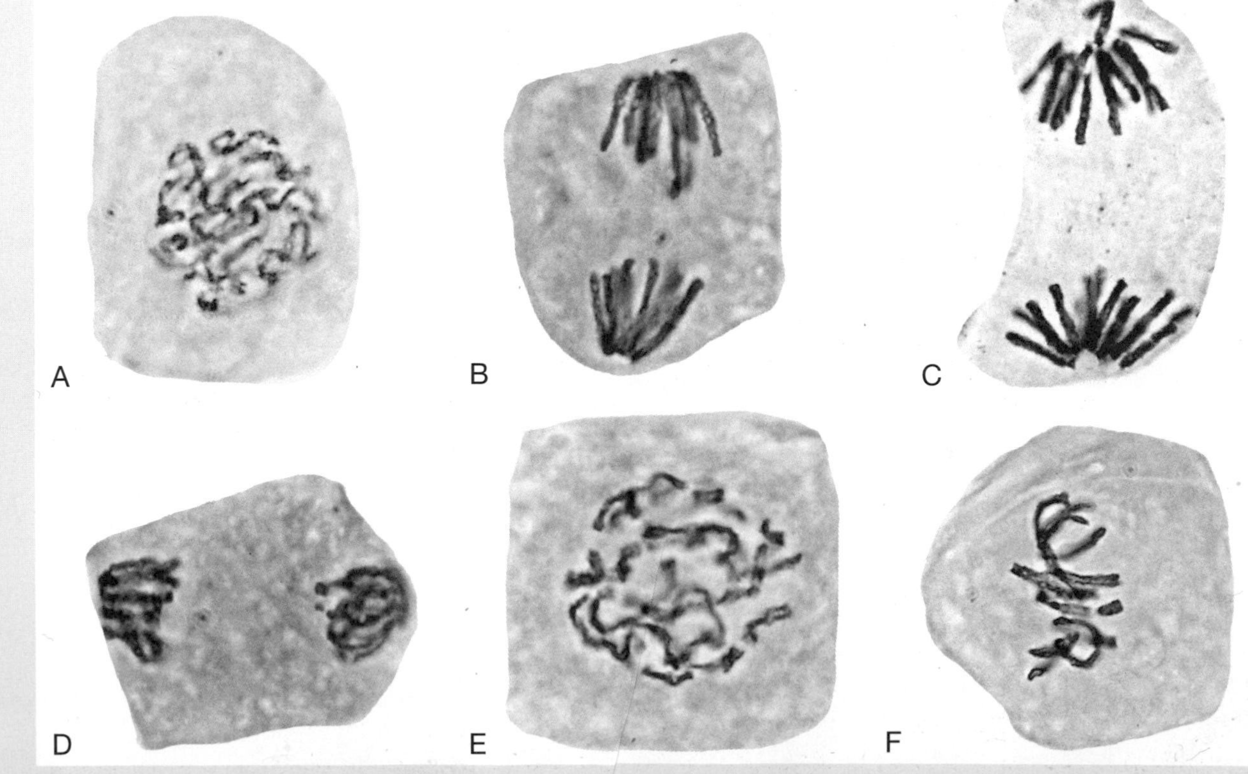

Answers on pages 42–43 **Answers** on pages 42–43 **Answers** on pages 42–43

(a) Name the stages of mitosis shown in the diagrams. **[6]**

(b) Explain the importance of mitosis to living organisms. **[3]**

[OCR Specimen]

---

3 Humans produce insulin from certain cells of the pancreas. The insulin gene is isolated from a human pancreas cell and then inserted into a plasmid. The DNA responsible for the synthesis of insulin is then inserted into a bacterium. The diagram, which is not drawn to scale, shows how insulin can be produced in this way. Different enzymes function at X and Y.

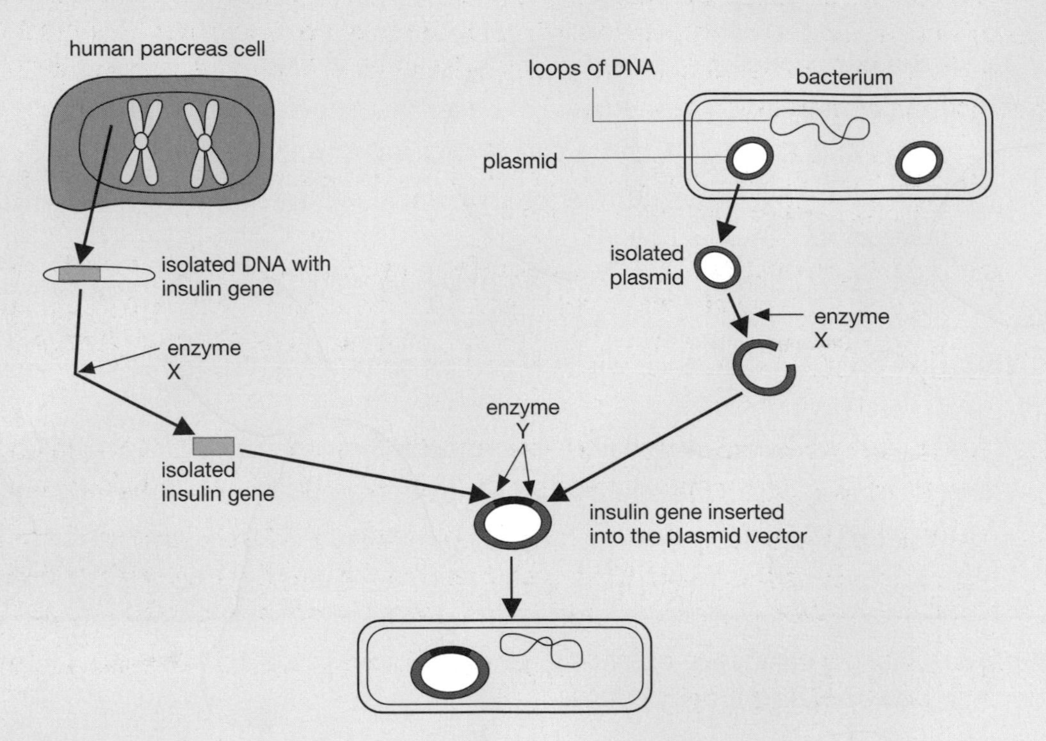

(a) State a general term for the technique shown in the diagram. **[1]**

(b) Outline the roles of the enzymes that function at X and Y. **[3]**

(c) Explain why the plasmid is described as a vector. **[2]**

(d) Outline the role of the bacterium in the process once the vector has been inserted into the host cell. **[4]**

[OCR Specimen]

Answers on pages 42–43 **Answers** on pages 42–43 **Answers** on pages 42–43

4 Some snakes produce poisonous venom. Different poisonous snakes produce different types of venom. A person who has been bitten by a particular species of poisonous snake may be treated with the appropriate antivenom. Antivenoms consist of antibodies against venom. They are made by injecting an animal such as a horse with increasing doses of snake venom. The antibodies the horse produces are then extracted and purified. 5

The Brazilian pit-viper is an extremely poisonous snake. The main component of its venom is jararhagin. Jararhagin is a protein which breaks down tissues, rapidly causing the death of any animal which has been bitten.

DNA technology may soon provide a better way of making antivenom. Instead of injecting animals with venom, they are injected with DNA. In one trial antivenom was produced by injecting DNA, coding for jararhagin, into cells in mice. The mice responded by producing antibodies to the jararhagin. 10

(a) **(i)** Describe how an animal, such as a horse, makes antibodies when it has been injected with snake venom (lines 4–5). [3]

(ii) When a person is bitten by a poisonous snake, doctors try to identify the snake so that the correct type of antivenom can be used. Explain why the bite of a particular species of poisonous snake must be treated with correct type of antivenom. [2]

(b) **(i)** The amino acid sequence of jararhagin is known. Explain how this information would enable a biologist to make an artificial gene which coded for jararhagin. [2]

(ii) The base sequence in this artificial gene may be different from the base sequence in the naturally occurring gene, even though they both code for the same protein. Use your knowledge of the genetic code to explain why. [2]

(iii) Describe how a cell from a mouse uses injected DNA to synthesise jararhagin protein (lines 11–13). [6]

[AQA A May 2002]

Answers on pages 42–43 **Answers** on pages 42–43 **Answers** on pages 42–43

Answers

(1) (a) **A** Phosphate **B** Pentose/sugar/deoxyribose **C** Bases/named bases

(b) **(i)** 8; **(ii)** 6; **(iii)** TACCG;

(c) Transcription;

(d) mRNA has codons, whereas each tRNA molecule has an anticodon.;

examiner's tip

If you cannot imagine the structure of each molecule try to draw out a quick diagram of each at the side of the paper. Detect the differences to score your marks.

---

(2) (a) A Prophase; B Anaphase; C Anaphase; D Late anaphase/early telophase; E Prophase; F Metaphase;

examiner's tip

LOOK FOR
Prophase: nucleus still visible, chromosomes have doubled.
Telophase: chromosomes have moved to the poles.
Anaphase: centromeres have just split, chromatids are beginning to move to the poles.
Late anaphase/early telophase: in transition between the two phases, chromatids moving to the poles.
Interphase: nucleus visible, chromosomes are not visible.
Metaphase: chromosomes are at the middle of the cell, i.e. the equator.

(b) The process produces genetically identical cells, for growth; for repair; for asexual reproduction.;

---

(3) (a) Genetic engineering/gene technology/gene manipulation/recombinant DNA technology;

examiner's tip

The recommendation here is to give genetic engineering. 'Mainstream' answers are the safest responses. All Examination Boards have difficult decisions to make when producing their mark schemes. Some answers are only just acceptable. Always try to give a mainstream answer.

(b) At X: cut the DNA open; cut at specific base sequences; cut to produce 'sticky' ends; (Any 1)
At Y: insulin gene attached to plasmid; to form a complete plasmid or ring of DNA; detail of recombination of DNA; (Any 2)

(c) The plasmid carries/transfers gene/DNA; to another cell/to another bacterium/to another place; (Any 2)

(d) The bacterium multiplies the plasmid or clones the plasmid; the bacterium reproduces by cloning/bacteria multiply; insulin is secreted/insulin is produced; bacterium uses chemicals produced by its own metabolism; (insulin) produced by protein synthesis.; (Any 4)

examiner's tip

After a gene has been transferred then the plasmid must be internally cloned. This is followed by the bacterium itself being cloned. This is necessary for large amounts of insulin to be produced.

(4) **(a)** **(i)** Venom contains antigens;
Macrophage/phagocyte presents antigen;
B lymphocytes become sensitised to specific antigens;
Clone/divide to form plasma cells;
Plasma cells produce specific antibody; (Any 3)

examiner's tip

There are key words here which you should include in your answer. Antigens in the snake venom stimulate the B lymphocytes to form plasma cells which produce the specific antibodies.

(ii) Different venoms contain different antigens;
Antibodies only fit with antigens of the right shape/complementary/specific;

examiner's tip

For each different antigen a specific antibody is produced. Take care with the words antigen and antibody. They are both involved in the immune response, so candidates often write down the incorrect term!

(b) **(i)** 3 bases in (RNA/DNA) code for one amino acid;
Work out correct base sequence;

examiner's tip

Each 3 bases are a codon so the sequence of codons along the coding strand can be copied.

(ii) Some amino acids are coded for by more than one codon/codon is degenerate;
Artificial gene may have different codons;
Natural gene may contain introns/junk DNA; (Any 2)

examiner's tip

Each amino acid can be coded for by up to six different codons! This fact is known as the degenerate code.

(iii)
1 DNA separates/hydrogen bonds break;
accept unzips NOT unwinds
2 To allow assembly of mRNA;
3 Using (m)RNA nucleotides;
4 Via RNA polymerase;
5 Complementary sequence or equivalent;
6 mRNA joins to ribosome;
allow travels to ribosome
7 tRNA carries a specific amino acid;
8 Codon-anticodon relationship/or explained/or defined;
9 Peptide bonds form between amino acids; (Any 6)

examiner's tip

The examiner wrote this question around the idea of the dangerous protein jararhagin, but the question is really only asking for your descripton of protein synthesis. Be ready for sequences like this! Once you can remember a sequence that you have learned, in the correct context, then the marks are easier to score.

Chapter 6

Continuity of life

Questions with model answers

C grade candidate – mark scored 4/6

1 (a) The diagram below shows a germinating pollen grain and a mature ovule from an insect pollinated flower. Some nuclei have been labelled.

For help see Revise AS Study Guide pages 96–99

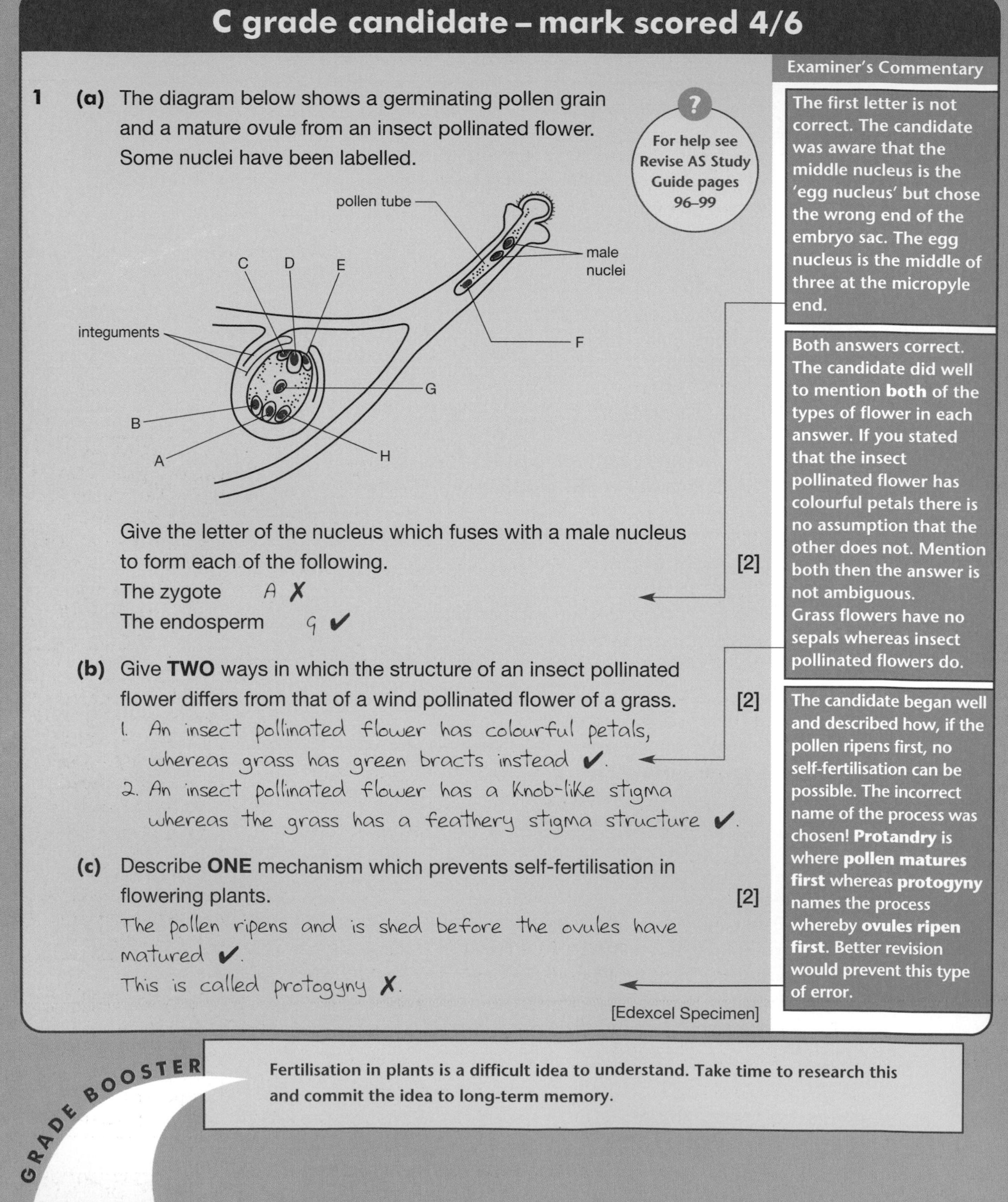

Give the letter of the nucleus which fuses with a male nucleus to form each of the following. [2]

The zygote A ✗

The endosperm G ✔

(b) Give **TWO** ways in which the structure of an insect pollinated flower differs from that of a wind pollinated flower of a grass. [2]

1. An insect pollinated flower has colourful petals, whereas grass has green bracts instead ✔.
2. An insect pollinated flower has a Knob-like stigma whereas the grass has a feathery stigma structure ✔.

(c) Describe **ONE** mechanism which prevents self-fertilisation in flowering plants. [2]

The pollen ripens and is shed before the ovules have matured ✔.

This is called protogyny ✗.

[Edexcel Specimen]

Examiner's Commentary

The first letter is not correct. The candidate was aware that the middle nucleus is the 'egg nucleus' but chose the wrong end of the embryo sac. The egg nucleus is the middle of three at the micropyle end.

Both answers correct. The candidate did well to mention **both** of the types of flower in each answer. If you stated that the insect pollinated flower has colourful petals there is no assumption that the other does not. Mention both then the answer is not ambiguous.
Grass flowers have no sepals whereas insect pollinated flowers do.

The candidate began well and described how, if the pollen ripens first, no self-fertilisation can be possible. The incorrect name of the process was chosen! **Protandry** is where **pollen matures first** whereas **protogyny** names the process whereby **ovules ripen first**. Better revision would prevent this type of error.

GRADE BOOSTER

Fertilisation in plants is a difficult idea to understand. Take time to research this and commit the idea to long-term memory.

A grade candidate – mark scored 15/15

2 The table shows the concentration of some sex hormones in the blood of a cow over a period of time.

Time/days	Concentration of hormone in the blood/arbitary units		
	progesterone	oestrogen	LH
0	1	14	32
2	2	8	1
4	4	7	1
6	10	7	1
8	14	7	1
10	18	7	1
12	19	7	1
14	19	7	1
16	18	7	1
18	8	18	1
20	1	14	32
22	1	8	32
24	2	8	1

(a) Use the figures in the table to estimate the length of the cow's oestrous cycle. Explain how you arrived at your answer. [2]

20 days ✔

I worked it out by noting that the progesterone level of 1 unit took 20 days to return to the same level. There seemed to be a pattern ✔.

(b) Explain how the high concentration of LH on day 0 caused an increase in progesterone in the days which followed. [3]

Luteinising hormone (LH) reached the ovary ✔, this stimulated ovulation ✔. After ovulation the follicle changes and can secrete progesterone ✔.

(c) Progesterone is responsible for the growth of the lining of the uterus and the development of its blood supply. Suggest how and explain why the figures for progesterone would differ from those in the table if the cow had become pregnant. [2]

The level of progesterone would increase ✔, because the lining of the uterus needs to be maintained through the pregnancy ✔.

Examiner's Commentary

For help see Revise AS Study Guide pages 103 and 104

Correct. Each hormone shows a pattern such as LH was at 32 units at time 0 days and it took 20 days to return to the identical level of 32 units. Any hormone could have been given as an example.

All correct. LH travels to the ovary via the blood, stimulates ovulation then the follicle changes into a **corpus luteum**. This latter part could also have gained credit but the candidate already scored a maximum by referring to the progesterone which it secretes.

The corpus luteum does not break down so quickly when the cow is pregnant. Additionally the placenta secretes a lot more progesterone.

(d) The concentration of progesterone in milk can be measured. It gives a very early indication of whether or not a cow is pregnant.

(i) Suggest how progesterone gets into milk. [1]

It may diffuse from the blood ✔.

(ii) Explain why it is an advantage for a farmer to know as early as possible whether or not a cow is pregnant. [2]

A cow needs to have another calf to maintain maximum milk flow ✔.

If the farmer finds out she is not pregnant then he will have her mated at next ovulation ✔.

(e) Describe how hormones may be used as contraception and in controlling infertility in humans. [5]

Oestrogen is produced which inhibits the production of the hormone FSH ✔ ✔.

In this way the Graafian follicle does not develop, so oestrogen acts as a contraceptive ✔.

If the person is infertile due to a lack of FSH then supplying it will stimulate the development of follicles, so ovulation will then take place ✔ ✔.

[AQA A Specimen]

Examiner's Commentary

All correct. In part (i) it is logical to suggest diffusion from the blood. All hormones are transported via blood! In answer to part (ii), an economic reason was needed. Milk volume decreases if no calf is produced, so the farmer aims for the production of a calf. Production of a calf, and the maintenance of milk are both economic reasons.

All correct! Progesterone could have been given but the candidate had already scored a maximum. Progesterone also inhibits FSH production and ensures that vaginal mucus is not the correct consistency for successful conception. The answers to this question part require continuous prose. Marking of the points would be based on the expression of logical points in clear scientific terms. The candidate would have also been credited with these communication marks.

Exam practice questions

1 The diagram below shows a section through a human ovary.

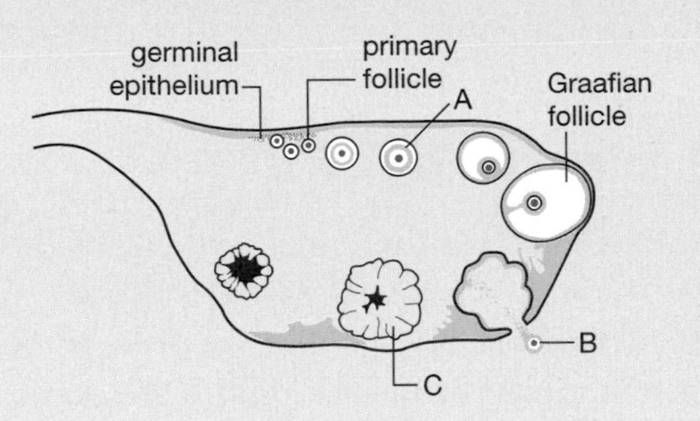

(a) Name parts A, B and C. [3]

(b) **(i)** Which part of the ovary divides to form the primary follicles? [1]

(ii) Which type of cell division is responsible for the production of the primary follicle? [1]

(c) FSH reaches the ovary so that part A begins to mature.

(i) Name the first hormone which is secreted by the ovary as a result of the arrival of FSH. [1]

(ii) Describe the role of this hormone in the menstrual cycle. [1]

(d) Structure B leaves the ovary.

(i) Where does structure B enter, immediately after leaving the ovary? [1]

(ii) Which hormone level peaks just before structure B leaves the ovary? [1]

---

2 The diagram below shows four gametes produced from a cell **X**. Only one chromosome is shown in each nucleus. Different combinations of two pairs of alleles are shown along these chromosomes. **A** is dominant to **a**, and **B** is dominant to **b**.

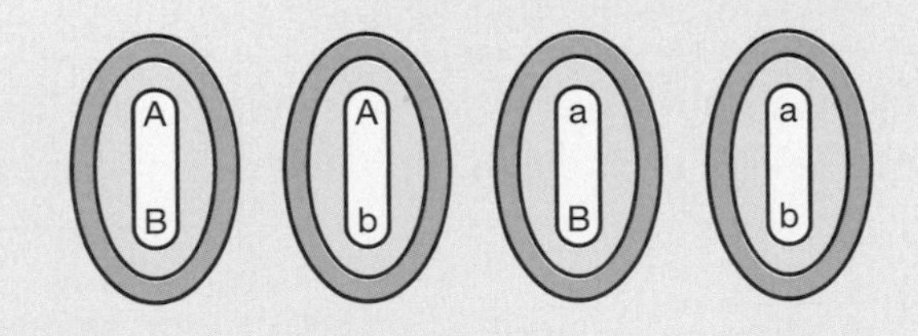

(a) Which type of cell division resulted in the production of these gametes from cell X? [1]

(b) Explain what, during this cell division, caused the production of the four *different* allele combinations. [2]

(c) Cell X divided to form the gametes shown above. Cell X had previously been produced as a result of two processes, fertilisation, then cell division.

- The parental cell which produced the male gamete was homozygous dominant.
- The other parental cell which produced the female gamete was homozygous recessive.
- These gametes fused to form cell X.

(i) Which type of cell division followed fertilisation? [1]

(ii) Draw a labelled diagram of cell X to show the allele combination along the chromosomes in the nucleus. Use the information given in parts **(a)** and **(c)** to help you. [3]

Answers on page 48 **Answers** on page 48 **Answers** on page 48

Answers

(1) (a) A Primary oocyte;
B Secondary oocyte;
C Corpus luteum;

(b) (i) Germinal epithelium;

(ii) Mitosis;

examiner's tip

The outer layer of cells (germinal epithelium) divides at the fetal stage to form oogonia. These go on to form primary follicles. Take care when giving answers about cell division. A key event in the ovary is meiosis but this only takes place as the primary oocyte divides. Before this stage divisions are mitotic!

(c) (i) Oestrogen;

(ii) It stimulates the build up of the endometrium/lining of the uterus.;

examiner's tip

When oestrogen is secreted into the blood the endometrium thickens and develops a capillary network in readiness for a potential embryo.

(d) (i) Fallopian tube/oviduct;

(ii) Luteinising hormone/LH;

examiner's tip

Fallopian tube should be easy to remember, but the hormone is more difficult. There are four important hormones controlling the menstrual cycle. Three of them were tested here.

---

(2) (a) Meiosis;

(b) Cross-overs/chiasmata;
and alleles are exchanged along the chromatids;

examiner's tip

Remember that the chromatids cross at the equivalent locus or position. Here the alleles can be exchanged from one chromatid to another.

(c) (i) Mitosis;

(ii) One chromosome has A + B;
One chromosome has a + b;
Two chromosomes in the nucleus;

Chapter 7 Energy and ecosystems

Questions with model answers

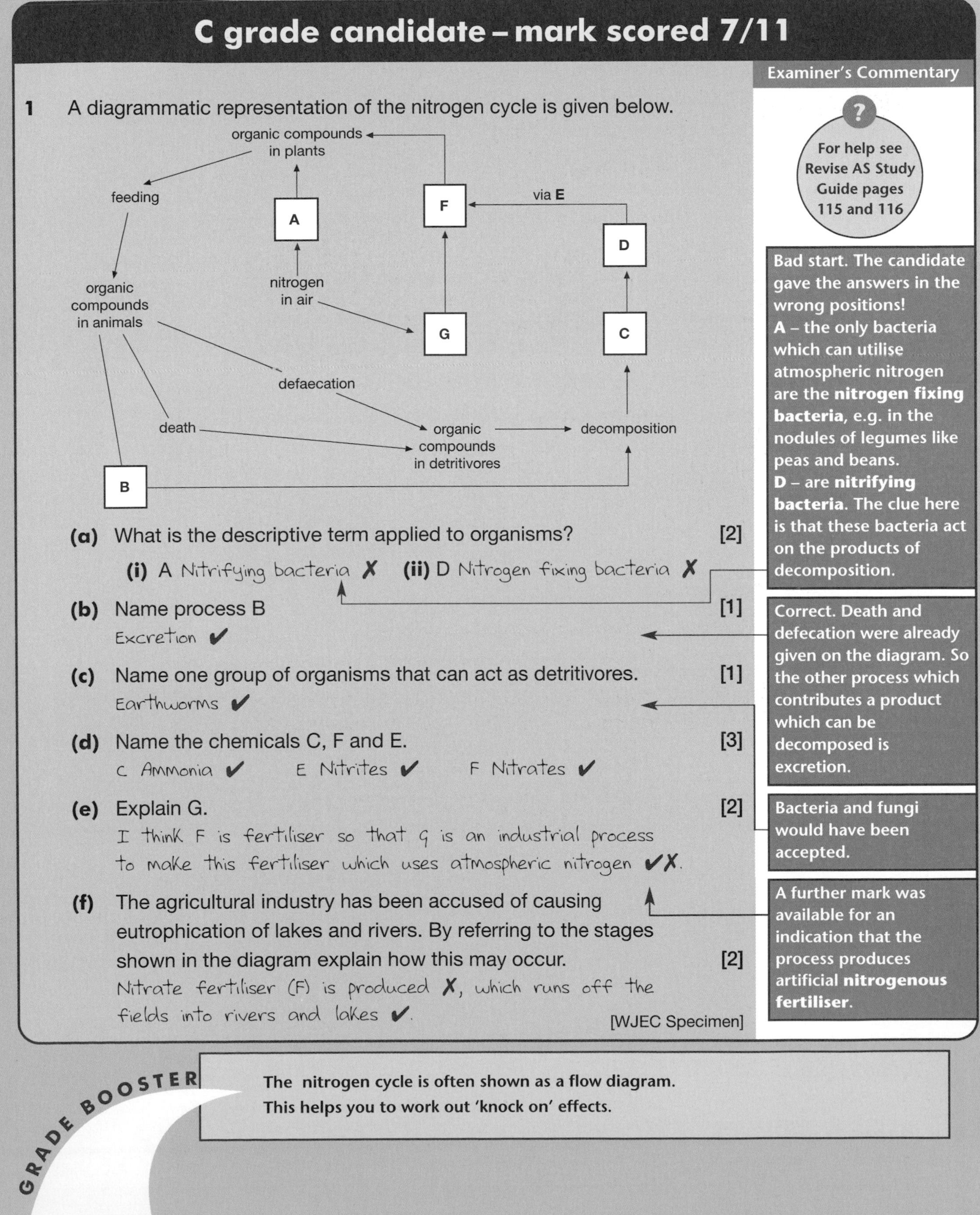

C grade candidate – mark scored 7/11

1 A diagrammatic representation of the nitrogen cycle is given below.

(a) What is the descriptive term applied to organisms? [2]

(i) A Nitrifying bacteria ✗ **(ii)** D Nitrogen fixing bacteria ✗

(b) Name process B [1]

Excretion ✔

(c) Name one group of organisms that can act as detritivores. [1]

Earthworms ✔

(d) Name the chemicals C, F and E. [3]

C Ammonia ✔ E Nitrites ✔ F Nitrates ✔

(e) Explain G. [2]

I think F is fertiliser so that G is an industrial process to make this fertiliser which uses atmospheric nitrogen ✔✗.

(f) The agricultural industry has been accused of causing eutrophication of lakes and rivers. By referring to the stages shown in the diagram explain how this may occur. [2]

Nitrate fertiliser (F) is produced ✗, which runs off the fields into rivers and lakes ✔.

[WJEC Specimen]

Examiner's Commentary

For help see Revise AS Study Guide pages 115 and 116

Bad start. The candidate gave the answers in the wrong positions!
A – the only bacteria which can utilise atmospheric nitrogen are the **nitrogen fixing bacteria**, e.g. in the nodules of legumes like peas and beans.
D – are **nitrifying bacteria**. The clue here is that these bacteria act on the products of decomposition.

Correct. Death and defecation were already given on the diagram. So the other process which contributes a product which can be decomposed is excretion.

Bacteria and fungi would have been accepted.

A further mark was available for an indication that the process produces artificial **nitrogenous fertiliser**.

GRADE BOOSTER

The nitrogen cycle is often shown as a flow diagram. This helps you to work out 'knock on' effects.

A grade candidate – mark scored 10/12

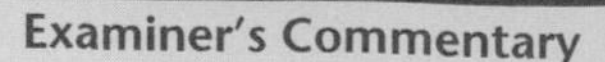

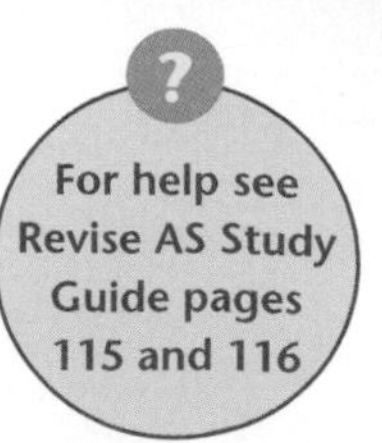

2 The population dynamics of the great tit (*Parus major*) in a wood near Oxford has been intensely studied over the past 50 years. Population changes relate to the production and survival of young great tits, and these are influenced by:

- the availability of insect prey (e.g. caterpillars) for adults and young
- the predation (e.g. by sparrowhawks) of young birds.

Adult great tits pair in autumn and establish territories in the following January. The females lay a clutch of about ten eggs in late April. These hatch over a period of two or three days, nearly two weeks after the last egg has been laid. Both adults catch caterpillars for the nestlings, which take two weeks to grow feathers. They then begin to make practice flights. They are fed by the adults for a further two weeks. The period just after the young have left the nest is critical and serious predation by sparrowhawks may take place.

Figures 1 to 4 below relate to information concerning the number of eggs produced, the survival of the nestlings (young in the nest) and the survival of the fledgelings (young to three months after leaving the nest).

Summarise the findings of each figure, and using the information provided previously, suggest an explanation for each finding.

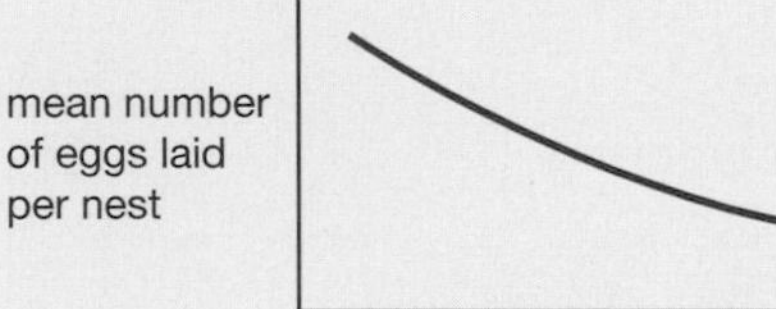

Figure 1 **[3]**

The average number of eggs per nest is greater, the fewer breeding pairs of birds there are in the wood ✔.
The less breeding pairs there are in the wood the more food there is for the birds so they can produce more eggs ✔. When the breeding pairs are close together, then more energy is expended for aggressive behaviour so that fewer eggs are laid ✔.

Analysing the graph, the higher the mean number of eggs the less breeding pairs there are in the area. So more food may be available per pair! All correct.

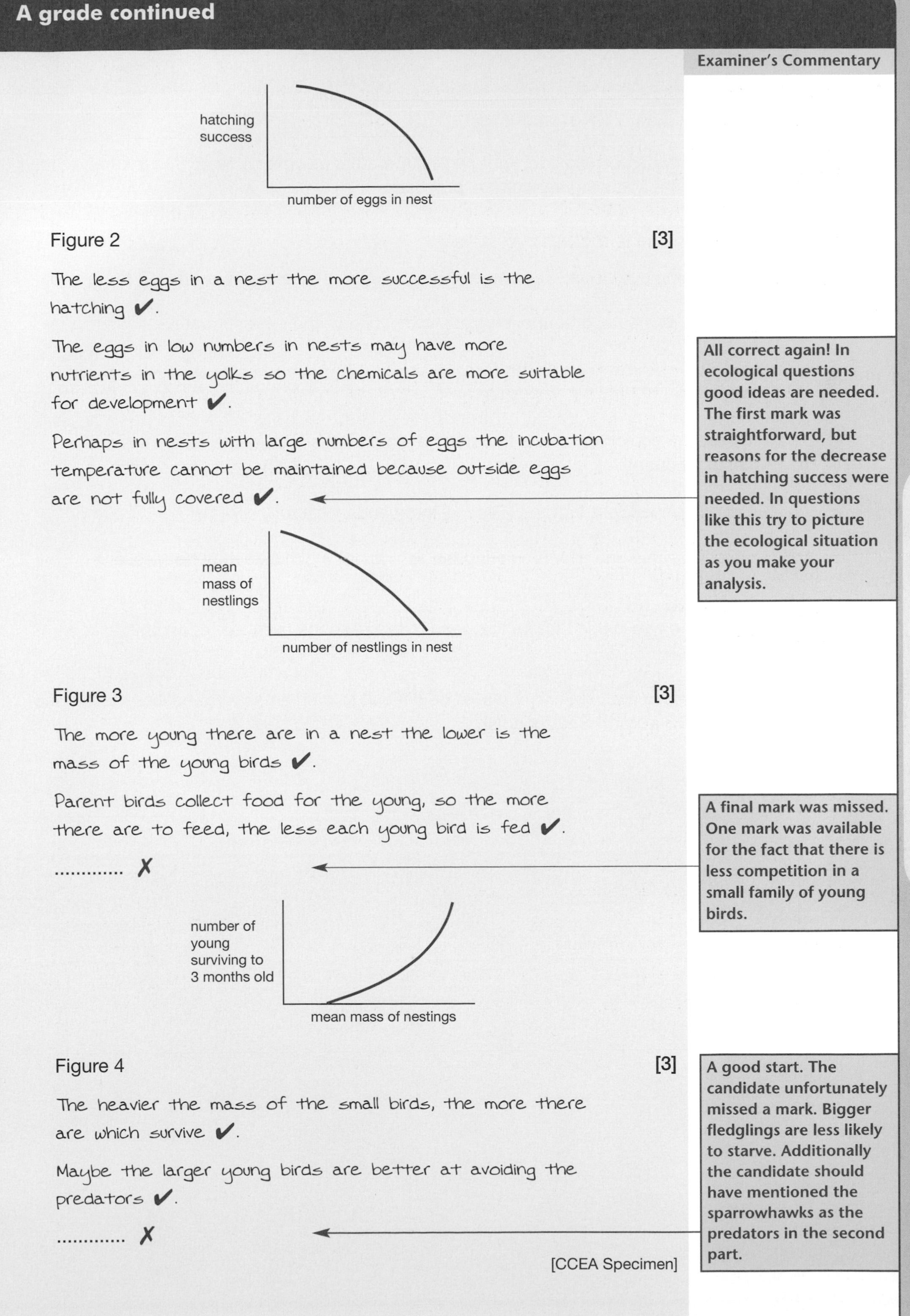

Examiner's Commentary

Figure 2 [3]

The less eggs in a nest the more successful is the hatching ✔.

The eggs in low numbers in nests may have more nutrients in the yolks so the chemicals are more suitable for development ✔.

Perhaps in nests with large numbers of eggs the incubation temperature cannot be maintained because outside eggs are not fully covered ✔.

All correct again! In ecological questions good ideas are needed. The first mark was straightforward, but reasons for the decrease in hatching success were needed. In questions like this try to picture the ecological situation as you make your analysis.

Figure 3 [3]

The more young there are in a nest the lower is the mass of the young birds ✔.

Parent birds collect food for the young, so the more there are to feed, the less each young bird is fed ✔.

.... ✗

A final mark was missed. One mark was available for the fact that there is less competition in a small family of young birds.

Figure 4 [3]

The heavier the mass of the small birds, the more there are which survive ✔.

Maybe the larger young birds are better at avoiding the predators ✔.

.... ✗

[CCEA Specimen]

A good start. The candidate unfortunately missed a mark. Bigger fledglings are less likely to starve. Additionally the candidate should have mentioned the sparrowhawks as the predators in the second part.

Exam practice questions

With respect to carbon dioxide, light and temperature as limiting factors in photosynthesis, comment on the following agricultural practices.

(a) Growing early crops under glass. [2]

(b) Introducing additional CO_2 into commercial greenhouses when growing tomatoes. [3]

(c) Painting domestic greenhouse glass with whitewash in the summer. [2]

[CCEA Specimen]

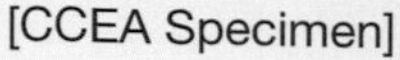

(a) **Circle** the ecological terms that can be used to describe human activities in the fishing industry.

Density dependent factor   Density independent factor   Parasite   Predator [2]

(b) State **two** reasons why regulating the fishing net mesh size may **not** prevent over-fishing. [2]

(c) Name the **two** biotic factors that tend to increase the numbers of an animal in a given area. [2]

(d) Suggest **two** reasons why it would be difficult to gain an accurate estimate of these factors for a North Sea food fish. [2]

[WJEC Specimen]

Answers on page 53 Answers on page 53 Answers on page 53

Answers

(1) (a) Low temperatures limit photosynthesis in the cold months in the UK climate.; Use of glass to cover plants, e.g. greenhouse warms up soil/or warms up atmosphere to increase the rate of photosynthesis.;

examiner's tip

An understanding of limiting factors is needed throughout this question. Glass always warms up the soil, above the ambient temperature. Photosynthetic rate is enhanced.

(b) An increase in CO_2 increases the rate of photosynthesis, over a range of different light intensities; and different summertime temperatures.;

examiner's tip

Gas heaters are often used which increase temperature and give out CO_2.

(c) The greenhouses can heat up excessively due to sunshine; light is not a limiting factor in the summer.;

examiner's tip

Whitewash enables the greenhouse to remain cooler by reflecting some of the light. If it was all reflected then there could be no photosynthesis.

---

(2) (a) Density independent factor; Predator;

examiner's tip

Humans have a predatory role as fish are caught for food. Density independent factor is correct. Fishing aims to maintain the population beneath the carrying capacity.

(b) The number of fishing boats in an area may not be regulated/a lot of fishing boats may be still be allowed to fish.; The time allowed for fishing may not be regulated.; Each species may have an optimum net mesh size.; Difficult to enforce. (Any 2)

examiner's tip

Always try to think logically around the problem. Net size is only one factor. The frequency of fishing and numbers of fishing vessels are both key factors.

(c) Birth rate; Immigration;

examiner's tip

Biotic factors are associated with the organisms in an area. The Examination Board may have credited two other responses. The numbers of predators in an area or the numbers of prey may influence the population of fish.

(d) Counting the number of females laying eggs is very difficult.; Counting the number of eggs which actually hatch is also a very difficult task.; Surveying the distribution of fish populations is very difficult as they are mobile.; Surveying the populations to find out the age structure of the population is very difficult.; The area is a huge size to survey.; (Any 2)

examiner's tip

All of the above factors could be applied to any biotic factor given in answer to part (c). Remember that in ecological type questions you will need lots of good logical ideas. At AS level you are expected to demonstrate analytical skills and apply them to new situations.

Chapter 8

Human health and disease

Questions with model answers

C grade candidate – mark scored 6/10

Examiner's Commentary

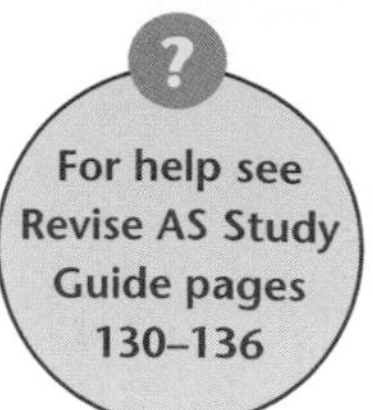

1 **(a)** Explain the meaning of the following terms as they apply to infectious diseases: [2]

Endemic – this means that the disease is always present in the population, such as malaria in the population of many African nations ✔

Epidemic – this is where a disease has spread throughout the population of an area, e.g. a city ✔

The table below shows the diseases which cause death in developing and developed countries.

developing countries		**developed countries**	
disease	**percentage deaths**	**disease**	**percentage deaths**
diarrhoea	32	heart diseases	32
respiratory infections: e.g. tuberculosis (TB)	25	cancers	23
malnutrition	10	strokes	12
malaria	7	bronchitis	6
measles	15	pneumonia	5
others	11	others	22

(b) With reference to the table above:

(i) Explain why infectious diseases are leading causes of death in developing countries. [4]

The food situation is not good ✗. They have poor living conditions ✗.

In the developing countries there may be no system of good sanitation so that diseases are spread via water ✔.

They cannot afford vaccinations ✔.

The first two responses did not justify a mark. The people often suffer from **malnutrition**. This reduces the effectiveness of the immune system. These are worthy of credit. Reference to the food situation not being good is much too vague. Similarly having poor living conditions is vague. What does this tell us? 'Poor living conditions encourage the spread of diseases' would have gained credit. Remember to give detail!

C grade continued

(ii) Explain why degenerative diseases are leading causes of death in developed countries. [4]

Many people in the developed countries are better off ✗.
A lot of people are overweight ✗.
The countries can afford to give the people vaccinations ✔.
Diseases like cancer take a longer time to develop and are not noticed until the disease is well established ✔.

[OCR Specimen]

Examiner's Commentary

The fact that developed countries have economic advantages is too vague. This could be linked to the fact that infectious diseases are controlled. 'Overweight' is another vague term. Obesity leading to a circulation problem, such as atherosclerosis or heart attack, gives the detail expected at this level.

GRADE BOOSTER

Many questions include data in the stem of the question. Be ready to analyse it logically and link the ideas to your specification (syllabus).

A grade candidate – mark scored 17/19

2 The bacterium, *Vibrio cholerae*, is the causative agent of cholera. The El Tor strain of *V. cholerae* originally occurred only in Indonesia. In 1961, this strain began to spread replacing existing strains in other parts of Asia. El Tor is now widespread throughout Asia, the Middle East, Africa and parts of Eastern Europe, but has never established itself in Western Europe.

El Tor is hardier than the strain it replaced and the bacteria may continue to appear in the faeces for up to three months after patients have recovered. The bacteria may persist in water for up to fourteen days.

(a) State **two** ways in which *V. cholerae* is transmitted from infected to unaffected people. [2]

1. It is carried by water, such as irrigation water on vegetables ✔.
2. Via faeces of an infected person ✔.

(b) Suggest how laboratory tests could identify carriers of cholera. [2]

Some people infected with cholera have mild symptoms or none at all, and are carriers of the disease.
Grow the bacteria from a sample of the person's faeces ✔.
Identify the bacteria ✔.

Examiner's Commentary

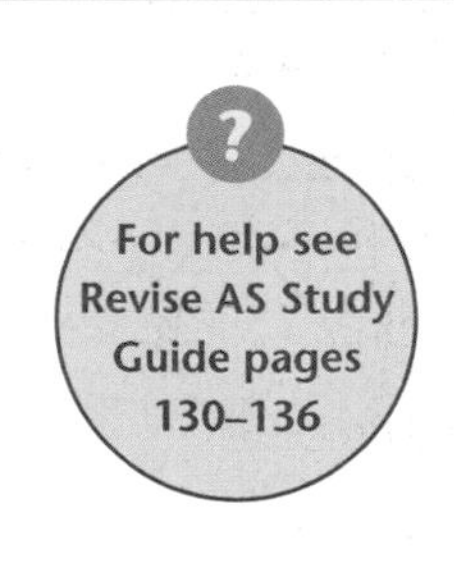

Both correct. Additional credit could have been given for transmission via food, drinking water or not washing hands after using the toilet or even carried by flies to food.

Correct but there were other possible answers. Test for antibodies against *V. cholerae*, use of a microscope, use of monoclonal antibodies.

Examiner's Commentary

(c) Suggest four reasons why El Tor has not become established in Western Europe. [4]

1. Sewage treatment takes place so that sanitation is good ✔.
2. So water is not contaminated ✔.
3. Water is purified with the help of chlorine to kill bacteria ✔.
4. .... ✗

> The candidate ran out of ideas after giving the first three correct responses. This example illustrates a good examination tip. A candidate could be tempted to give one answer about water, but here two are credited. Give full detail to gain full credit.
> The Examination Board **rejected** vaccination or use of antibiotics.

The United Nations, recognising that most of the outbreaks of cholera were the result of polluted water supplies, set up a 'Decade of Water' in 1981. Its aim was to provide safe water for everyone. Over the decade 1981/1990, the number of people lacking a safe water supply in developing countries dropped from 1800 million to 1200 million.

(d) Explain why cholera continues to be a worldwide problem, in spite of the 'Decade of Water' campaign. [8]
(In this question, 1 mark is available for the quality of written communication.)

The cholera bacterium is passed via polluted water.
Many people still do not have a supply of clean water ✔.
Only 600 million people out of the 1800 million people were given access to a safe water supply, leaving 1200 million people vulnerable ✔. These people did not have access to clean water ✔. The programme is not 100% effective. In fact it is around 66.6% effective ✔.
One of the problems may be that areas are large and rural. They are difficult to reach and supply with the safe water supplies ✔. Additionally, the population in the area is increasing, so there is more chance of the diseases being passed on ✔.
Suitable housing is not always available so that shanty towns develop with poor sanitation ✔.

(Communication mark) ✔

> Full marks would be awarded. The last sentence has two valid points, i.e. the build up of shanty towns, and poor sanitation. The candidate reached the maximum for this part of the question and could not gain further credit.
> A further mark would have been credited for **written communication**. The answer satisfied these criteria, 'legible text with accurate spelling, punctuation and grammar'.

The antibiotic tetracycline is sometimes used as a treatment for cholera.

(e) **(i)** Suggest **two** ways in which tetracycline can affect *V. cholerae*. [2]

It is bacteriostatic, preventing further growth of bacterial cells ✔.
Specifically it prevents protein synthesis ✔.

> Full marks, but others were available, such as the tetracycline causing the lysis or splitting open of the cell membrane.

(ii) Explain why tetracycline should not be used routinely for all cases of cholera. [1]

Bacteria become immune to the antibiotic ✗.

[OCR Specimen]

> The candidate is close, but would not be awarded credit. The bacteria can become **resistant** to the antibiotic. Immune is not an equivalent term.

Exam practice questions

1 **(a)** What is a disease? [1]

(b) Describe FOUR different types of disease. Give a specific example of each type of disease in your answer. [8]

(b) What is the meaning of the term pandemic? [1]

---

2 **(a)** Read the following passage on the disease malaria, and then write in an appropriate word or words on each dotted line to complete the passage.

The disease, malaria, is caused by the protozoan parasite, ... This dangerous parasite is carried by a .... This is a .... mosquito. The mosquito may feed on a person who is suffering from malaria. It does this at night by inserting its .... into a .... beneath the skin. The digestion of the mosquito releases the malarial parasites which burrow into the insect's stomach wall and breed there. Some move to the .... Next time the mosquito feeds it secretes saliva to prevent clotting of the blood. This introduces the .... into the person's blood, who is likely to contract the disease. [8]

(b) Describe FIVE different ways of preventing malaria. [5]

Answers on page 58 **Answers** on pages58 **Answers** on page 58

Answers

(1) (a) A disease is a disorder of a tissue, organ or system of an organism. As a result of the disorder symptoms are evident.;

(b) **Infectious disease by pathogens** which attack an organism; and can be passed from one person to another, e.g. infectious disease to include measles, etc.; **Genetic diseases** (congenital diseases) can be passed from parent to offspring; e.g. any genetic disease to include haemophilia, cystic fibrosis, etc.;

Dietary related diseases, as a result of the foods which we eat.; Too much or too little food may cause disorders, e.g. obesity/anorexia nervosa/deficiency diseases, e.g. lack of vitamin D causes the bone disease rickets.;

Environmentally caused diseases where some aspect of the environment disrupts bodily processes; e.g. as a result of nuclear radiation leakage, cancer may result.;

Auto-immune disease where the body, in some way attacks its own cells, so that processes fail to function effectively, e.g. leukaemia where phagocytes destroy a person's own red blood cells.

(Any 4 diseases)

(c) This refers to an outbreak of a disease over a very large area, such as a continent.;

---

(2) (a) The disease, malaria, is caused by the protozoan parasite, **Plasmodium**.; This dangerous parasite is carried by a **vector**.; This is a **female**; **Anopheles**; mosquito. The mosquito may feed on a person who is suffering from malaria. It does this at night by inserting its **stylet**; into a **blood vessel**; beneath the skin. The digestion of the mosquito releases the malarial parasites which burrow into the insect's stomach wall and breed there. Some move to the **salivary glands**.; Next time the mosquito feeds it secretes saliva to prevent clotting of the blood. This introduces the **parasites**; into the person's blood, who is likely to contract the disease.

examiner's tip

This is known as a cloze exercise. Knowledge is required but the clues are there to help you. Logically the first word is Plasmodium, the correct term for the malarial parasite. The tricky marks are the two given for female, and Anopheles mosquito. The key skill here is to remember the detail.

(b) Insecticide sprayed onto lake surfaces kills mosquito larvae/oil poured on lake surface prevents air entering the breathing tubes of the (mosquito) larvae so they die.

Fish can be used as predators in lakes to eat the larvae/use biological control.

Drain ponds to remove the mosquitoes' breeding area/cover up all waste containers to remove the breeding area.

Use *Bacillis thuringiensis* to destroy mosquitoes.

Mosquito nets exclude mosquitoes from buildings/from beds.

Electronic insect killer attracts the mosquitoes via ultra-violet light then kills them by application of voltage.

Use of drugs to kill the malaria organism in sufferers/isolate the people suffering from malaria.

(Any 5)

AS Mock Exam 1

Centre number ______________________
Candidate number ______________________
Surname and initials ______________________

Examining Group

Biology

Time: 1 hour 25 minutes **Maximum marks: 71**

Answer **all** questions.

Section A

1 **(a)** The diagram shows a human heart.

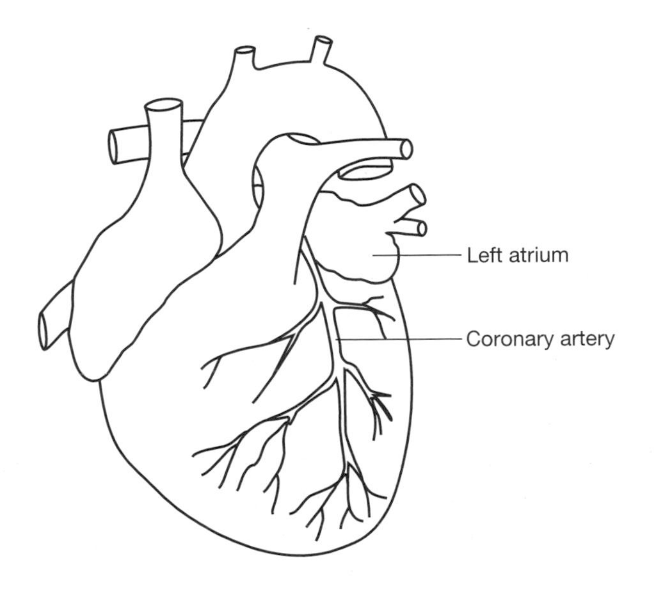

(i) Use a guideline and label to show the position of the sinoatrial nose (SAN) **[1]**

(ii) What is the function of the coronary artery?

....

.... **[1]**

(b) Impulses spread through the walls of the heart from the SAN. The table shows the rate of conduction of impulses through various parts of the conducting tissues.

Part of pathway	**Rate of conduction/ms^{-1}**	**Mean distance/mm**
From SAN to atrioventricular node (AVN) across atrium	1.0	40
Through AVN	0.05	5
From AVN to lower end of bundle of His	1.0	10
Along Purkyne fibres in ventricle walls	4.0	–

(i) Calculate the mean time taken for an impulse to pass from the SAN to the lower end of the bundle of His. Show working.

.... s **[2]**

(ii) Explain the advantage of the slow rate of conduction through the AVN.

....

....

....

.... **[2]**

(iii) Suggest **one** advantage of the high rate of conduction in the Purkyne fibres which carry impulses through the walls of the ventricles.

....

.... **[1]**

(c) How would cutting the nerve connections from the brain to the SAN affect the beating of the heart?

....

.... **[1]**

[8 marks]

[AQA B 2003]

2 The graphs below illustrate the results of two experiments in which the activity of a single enzyme was investigated over a range of temperatures. In one experiment, the enzyme was in its soluble form. In the other experiment, an identical concentration of the same enzyme was in an immobilised form. Study the graphs and use them to help you answer the questions which follow.

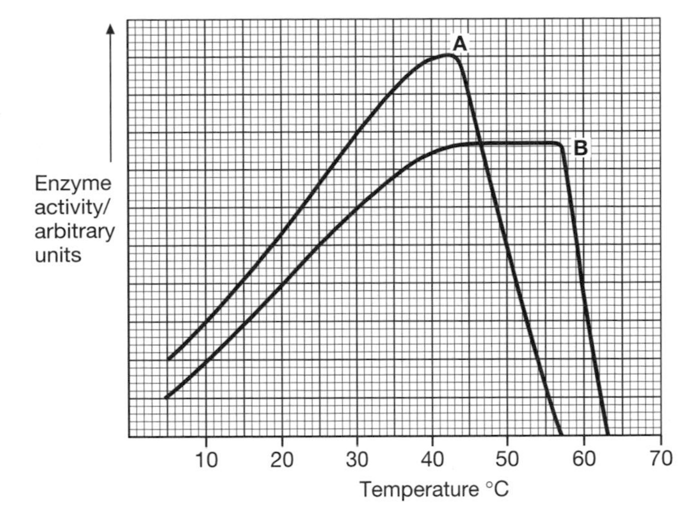

(a) In graph **A**, at which temperature does this enzyme achieve its maximum activity.

.... **[1]**

(b) At which temperature are both forms of this enzyme equally active.

.... **[1]**

(c) **(i)** Which of the two graphs **A** or **B**, represents the results for the enzyme in immobilised form?

....

....

.... **[1]**

(ii) Identify, and explain, **two** pieces of evidence in the graph that justify your decision in part **(i)**.

1. ....

....

2. ....

.... **[2]**

(d) Explain the rapid decline in activity of both forms of the enzyme at the higher temperatures investigated.

....

.... **[1]**

[6 marks]

[CCEA, January 2003]

3 The figure shows the flow of energy through the trees in a forest ecosystem. The numbers represent inputs and outputs of energy in kilojoules per m^2 per year.

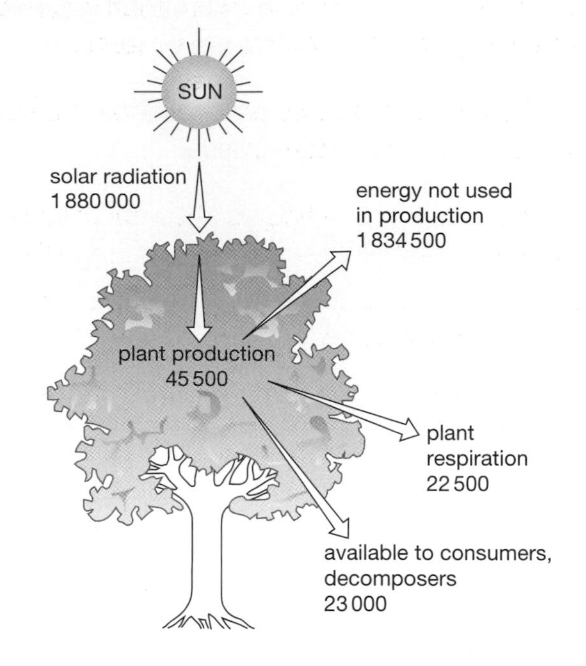

(a) (i) On the figure, draw a ring around the number which indicates the energy entering the system via photosynthesis. **[1]**

(ii) The total energy available to the plant in the ecosystem is 1 880 000 kJ per m^2 per year.

Calculate the efficiency of photosynthesis. Show your working.

Efficiency = ... **[2]**

(b) Suggest **four** reasons why so much solar energy is **not** used in the production in the forest ecosystem.

1. ....

2. ....

3. ....

4. .... **[4]**

(c) In what form will energy from plant respiration escape from the ecosystem?

.... **[1]**

[8 marks]

[OCR Specimen]

4 An experiment was carried out to determine what happens to amino acids after they are absorbed by animal cells. The cells were incubated for 5 minutes in a medium containing radioactively labelled amino acids. The radioactive amino acids were then washed off and the cells were incubated in a medium containing non-radioactive amino acids.

Samples of the cells were taken at 5, 10 and 45 minutes after the start of the experiment and the sites of the radioactivity in the cells were determined.

The results are given in the table below. The figures show radioactivity in certain cell organelles expressed as a percentage of the total radioactivity within the cells.

Organelle	Percentage of total radioactivity		
	At 5 minutes	At 10 minutes	At 45 minutes
Rough endoplasmic reticulum	80	10	5
Golgi apparatus	10	80	30
Secretory vesicles	0	5	60

(a) Name ONE type of molecule synthesised from amino acids in cells.

.... [1]

(b) Explain why the radioactivity is associated mainly with the rough endoplasmic reticulum after the first 5 minutes of the experiment.

....

....

.... [2]

(c) Explain the changes in the pattern of radioactivity in the cell during the remaining 40 minutes of the experiment.

....

....

.... [3]

(d) Suggest why the figures in the table total less than 100%.

....

....

.... [2]

(e) If the experiment is continued for a further period of time, most of the radioactivity will be found outside the cell.

Name and describe the process which brings about this result.

....

....

.... [3]

[11 marks]

[Edexcel Specimen]

Section B

1 Read the following passage.

Forensic science has come a long way from the magnifying glass and deerstalker image of Sherlock Holmes. Take the identification of blood stains as an example. There has been a change in the sensitivity of tests used to distinguish between blood obtained from different individuals. Early techniques relied on methods based on the biology of red blood cells; more modern ones are based on white blood cells. 5

Until relatively recently all that a forensic scientist could do was identify the blood groups concerned. Certain protein molecules in the blood act as antigens and it is the presence of these which determines blood group. The four blood groups A, B, AB and O are determined by the presence of the relevant antigen. For example, individuals with blood group A, have antigen A, while those with blood group B have antigen B. 10

More modern techniques have allowed us to progress much further. Instead of looking at a rather limited range of proteins we can now look at DNA itself. Genetic fingerprinting can be used to distinguish between individuals by looking at similarities and differences in part of their DNA. Some of the non-coding DNA consists of short sequences of bases which may be repeated. The actual number of times these sequences are repeated varies from individual to individual. 15 Genetic fingerprinting compares these sequences. The flow chart summarises the main steps involved in the procedure.

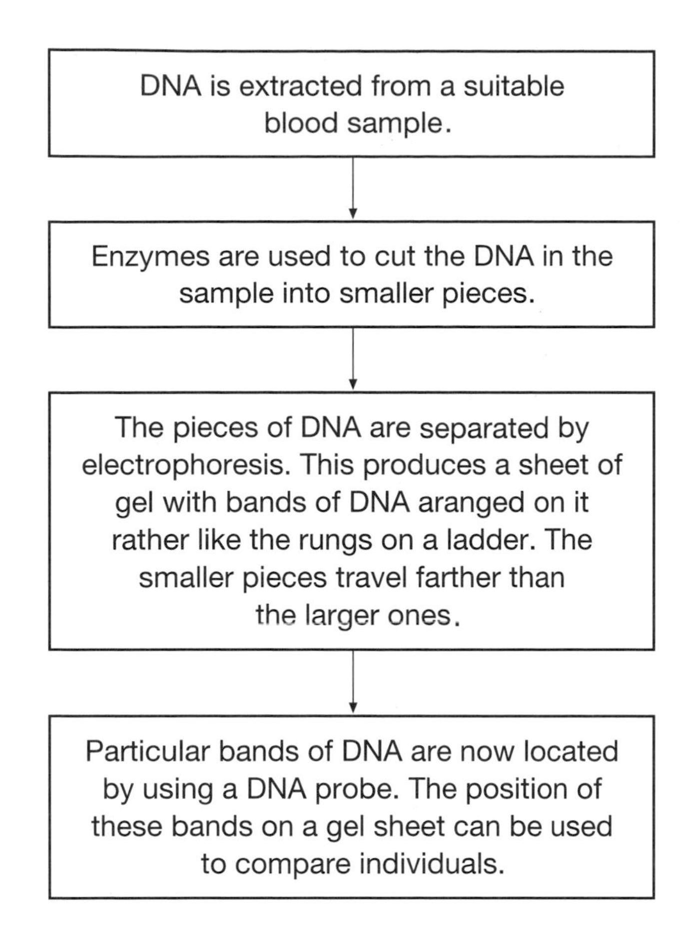

Use information from the passage and your own knowledge to answer the following questions.

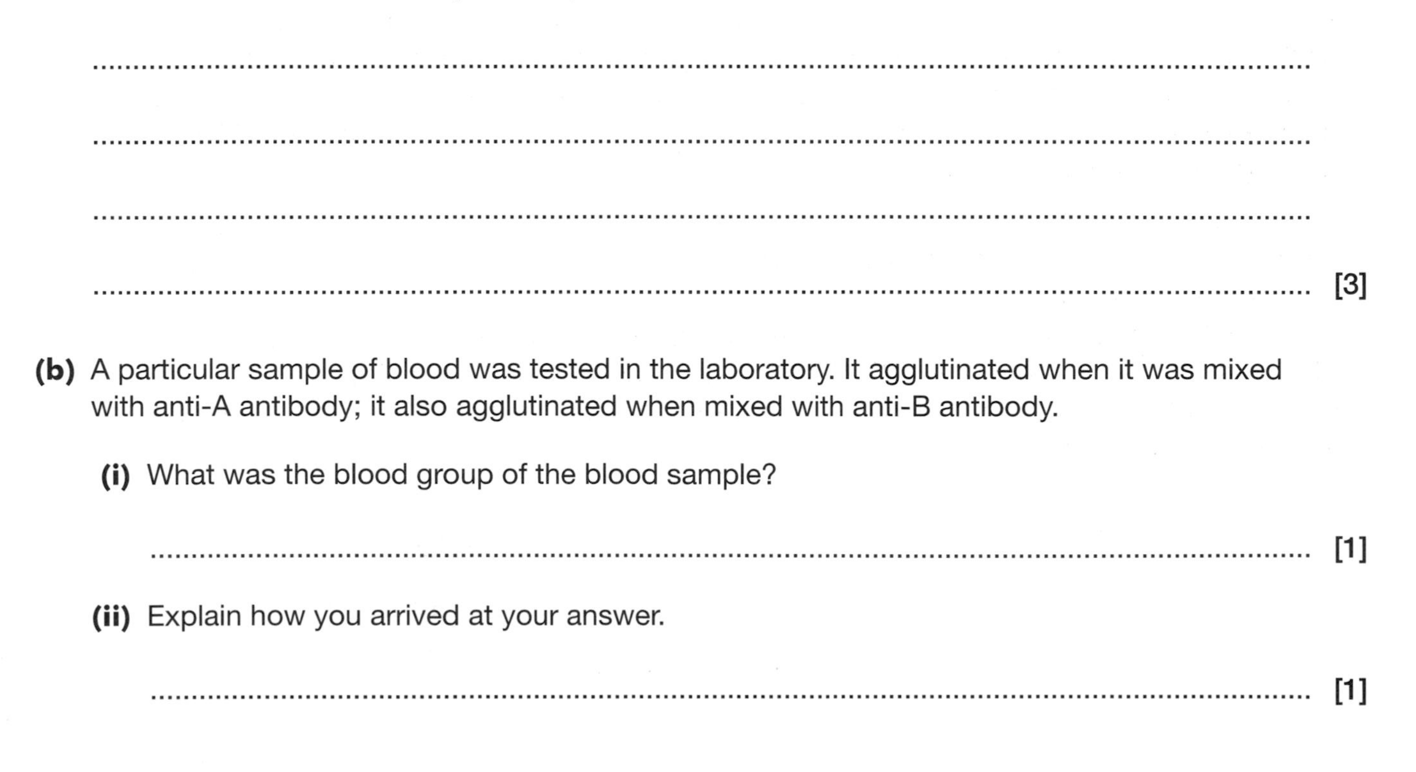

(a) Explain why early techniques used to distinguish between blood obtained from different individuals relied on methods based on the biology of red cells while more modern ones are based on white blood cells (lines 4–5).

....

....

....

.... [3]

(b) A particular sample of blood was tested in the laboratory. It agglutinated when it was mixed with anti-A antibody; it also agglutinated when mixed with anti-B antibody.

(i) What was the blood group of the blood sample?

.... [1]

(ii) Explain how you arrived at your answer.

.... [1]

(c) Explain what is meant by 'non-coding' DNA (line 14).

....

.... [1]

(d) **(i)** Name the type of enzyme used to cut DNA in the sample into smaller pieces (box 2).

.... [1]

(ii) Explain why the lengths of pieces produced by cutting the DNA with one of these enzymes will vary from individual to individual.

....

....

....

....

....

.... [3]

(e) Describe how a DNA probe may be used to find a particular band of DNA.

....

....

....

....

....

....

....

....

.... [5]

[15 marks]

[AQA A Specimen]

2 A procedure was carried out to separate the major organelles within liver cells. This involved breaking up (homogenising) liver tissue in an ice cold salt solution which had the same water potential as the cell cytoplasm.

Ultracentrifugation was then used to separate the organelles. Ultracentrifugation is a process that separates materials of different densities by spinning them in a tube at different speeds. The denser materials are forced to the bottom of the tube as a pellet, while the less dense materials remain nearer the top of the tube in a liquid known as the supernatant.

The flow chart below summarises the steps involved in this procedure.

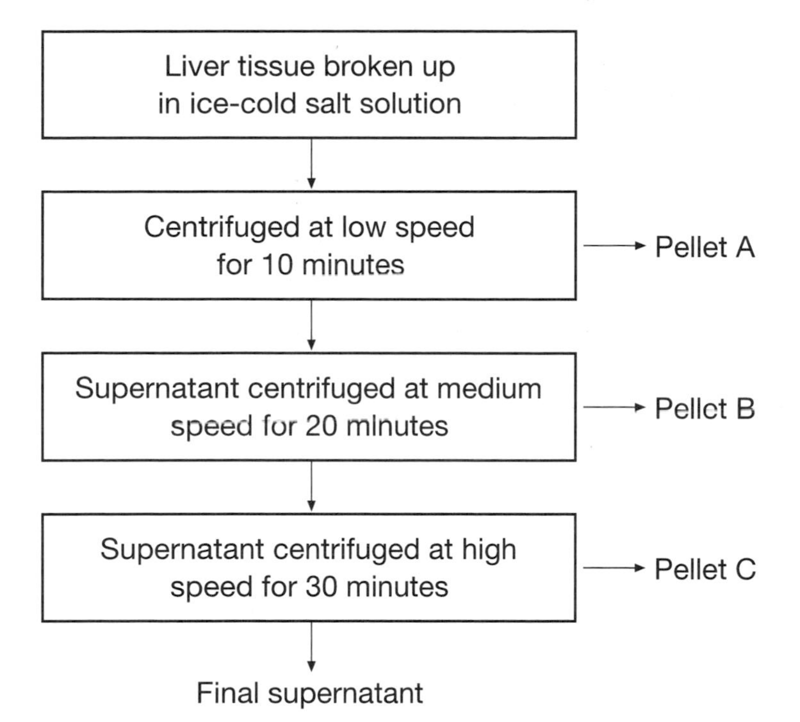

(a) Suggest why it was necessary for the salt solution to have the same water potential as the cell cytoplasm.

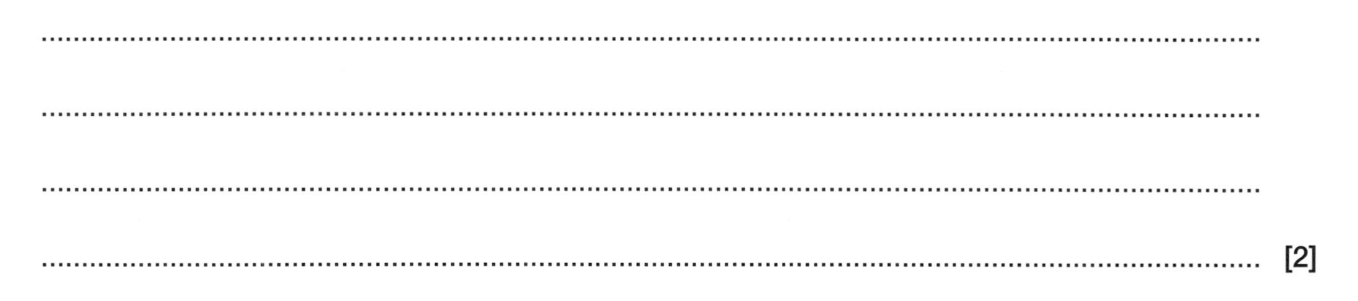

[2]

(b) This procedure separated mitochondria, nuclei and ribosomes into the three pellets A, B and C shown on the flow chart. Complete the table below to show which one of these three organelles would be found in which pellet.

Pellet	Organelle
A	
B	
C	

[3]

(c) Suggest **two** components of the cell, other than water, that might be present in the final supernatant.

1 ....

2 .... [2]

(d) In the space below, draw and label a diagram to show the structure of a mitochondrion.

[4]

(e) Explain why large numbers of mitochondria are found in liver cells.

....

....

....

.... [2]

[13 marks]

[Edexcel Jan 2002]

3 Answer **ONE** of the following questions.

Either – (a) Explain what is meant by translation and describe how the cell carries out translation.

Or – (b) Describe the **changes** that occur in a cell dividing by meiosis from the point at which the chromosomes first become visible under the light microscope.

....

....

....

....

....

....

....

....

....

....

....

....

....

....

....

....

....

....

....

....

.... [10]

[WJEC Specimen]

AS Mock Exam 2

Centre number ____________________

Candidate number ____________________

Surname and initials ____________________

Examining Group

Biology

Time: 1 hour 35 minutes   Maximum marks: 76

Answer **all** questions.

---

1 The diagram shows an epithelium found in a mammal.

X

A

B

Y

× 800

(a) State the type of microscope used to view the epithelium preparation.

.... [1]

(b) State the parts labelled **X** and **Y** on the diagram.

X ....

Y .... [2]

(c) Such cells are found in the trachea. Explain how the structure of the cells is related to their function.

....

....

.... [2]

(d) Calculate the actual height of the cell shown between the lines labelled **A** and **B**. Show your working.

....

.... [2]

(e) State what is meant by the term tissue.

....

.... [2]

[9 marks]

[WJEC A 2003]

2 The diagram below shows the plasma membrane of an animal cell.

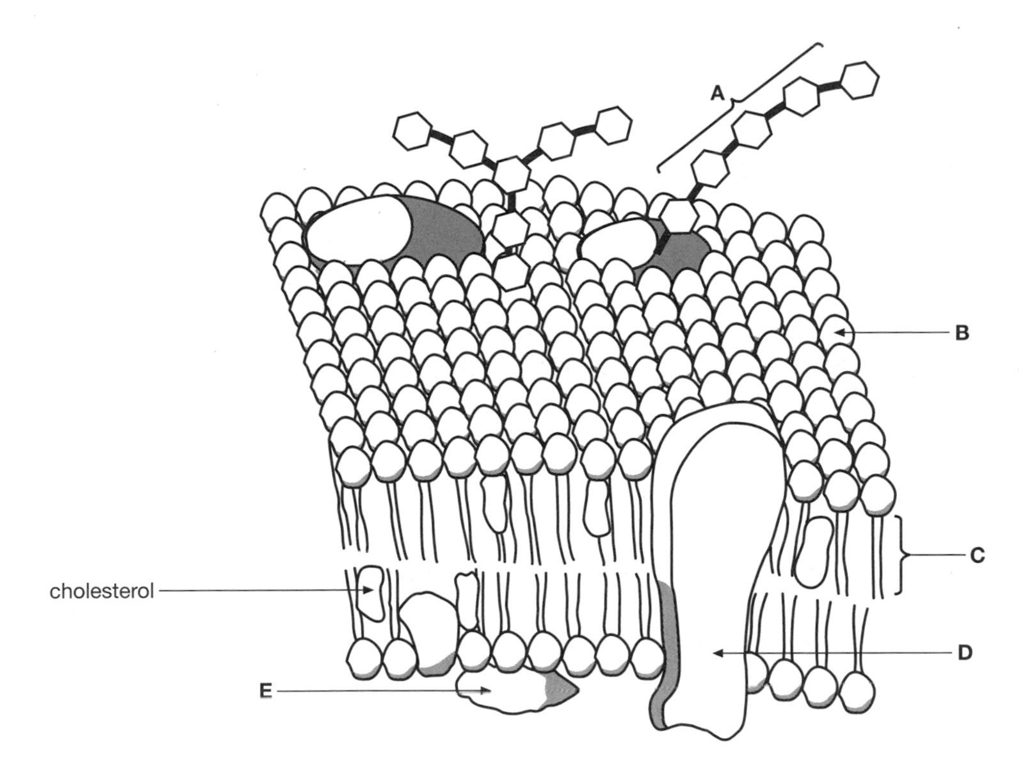

With reference to the diagram:

(a) **(i)** State the names of the structures A to E.

A ....

B ....

C ....

D ....

E .... **[5]**

(ii) State the name given to this model of membrane structure.

.... **[1]**

(b) Glucose and vitamin A are molecules that enter a cell by passing across the membrane. Glucose is water soluble and vitamin A is fat soluble.

Explain how the properties of the molecules and the structure of the membrane determine the way in which these two molecules pass across.

Glucose

....

....

.... [2]

Vitamin A

....

....

.... [2]

(c) Name **two other** methods by which substances cross the plasma membrane.

.... [2]

[12 marks]

[WJEC A 2003]

3 **(a)** The diagram shows a stage of mitosis in an animal cell.

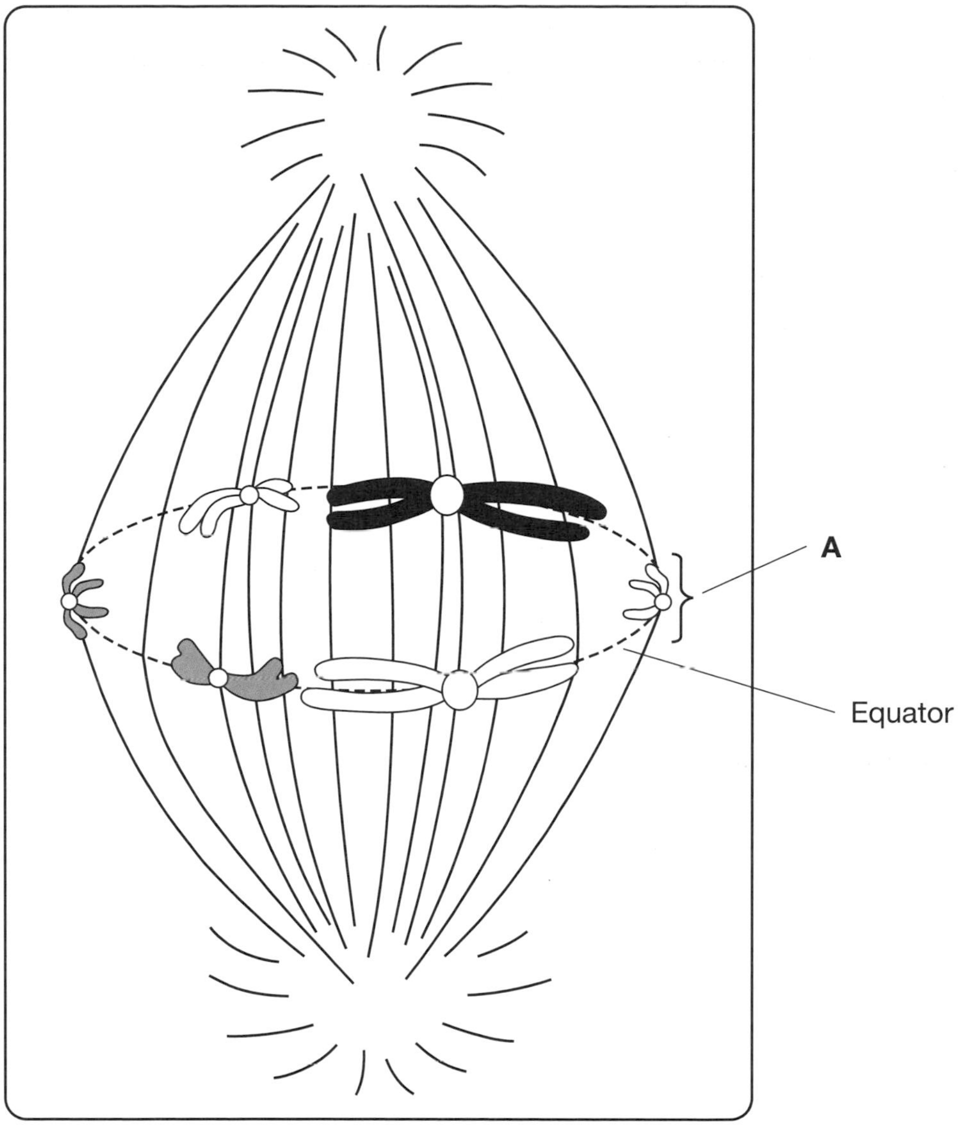

A

Equator

(i) Name this stage of mitosis.

.... **[1]**

(ii) What happens to structure **A** in the stage immediately following that shown in the diagram?

....

....

....

.... **[2]**

(iii) Complete the diagram to show the chromosomes for a gamete produced by this animal.

[2]

(b) The amount of DNA in cells from a tissue undergoing mitosis was analysed. Some cells were found to have 7.6 units of DNA, others had only 3.8 units. Explain why.

....

....

....

.... **[2]**

[7 marks]

[AQA B 2003]

4 The diagrams below show three different types of blood vessels. They are not drawn to the same scale.

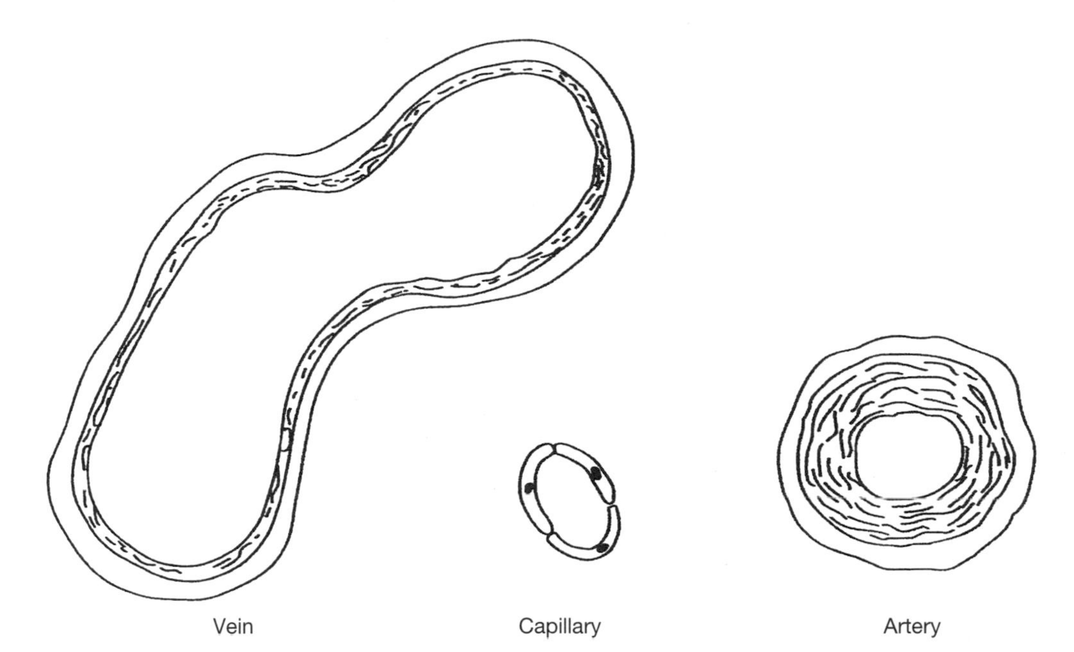

(a) Give a characteristic of each blood vessel and explain how it is related to its function.

Vein ....

....

....

....

Capillary ....

....

....

....

Artery ....

....

....

.... **[6]**

(b) The following diagram shows blood pressure changes as blood travels through one circuit of the circulatory system.

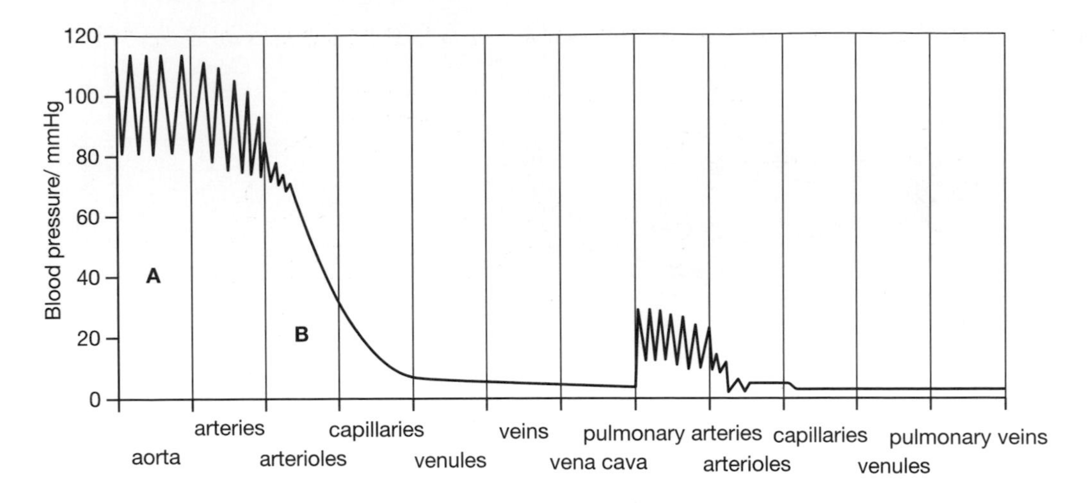

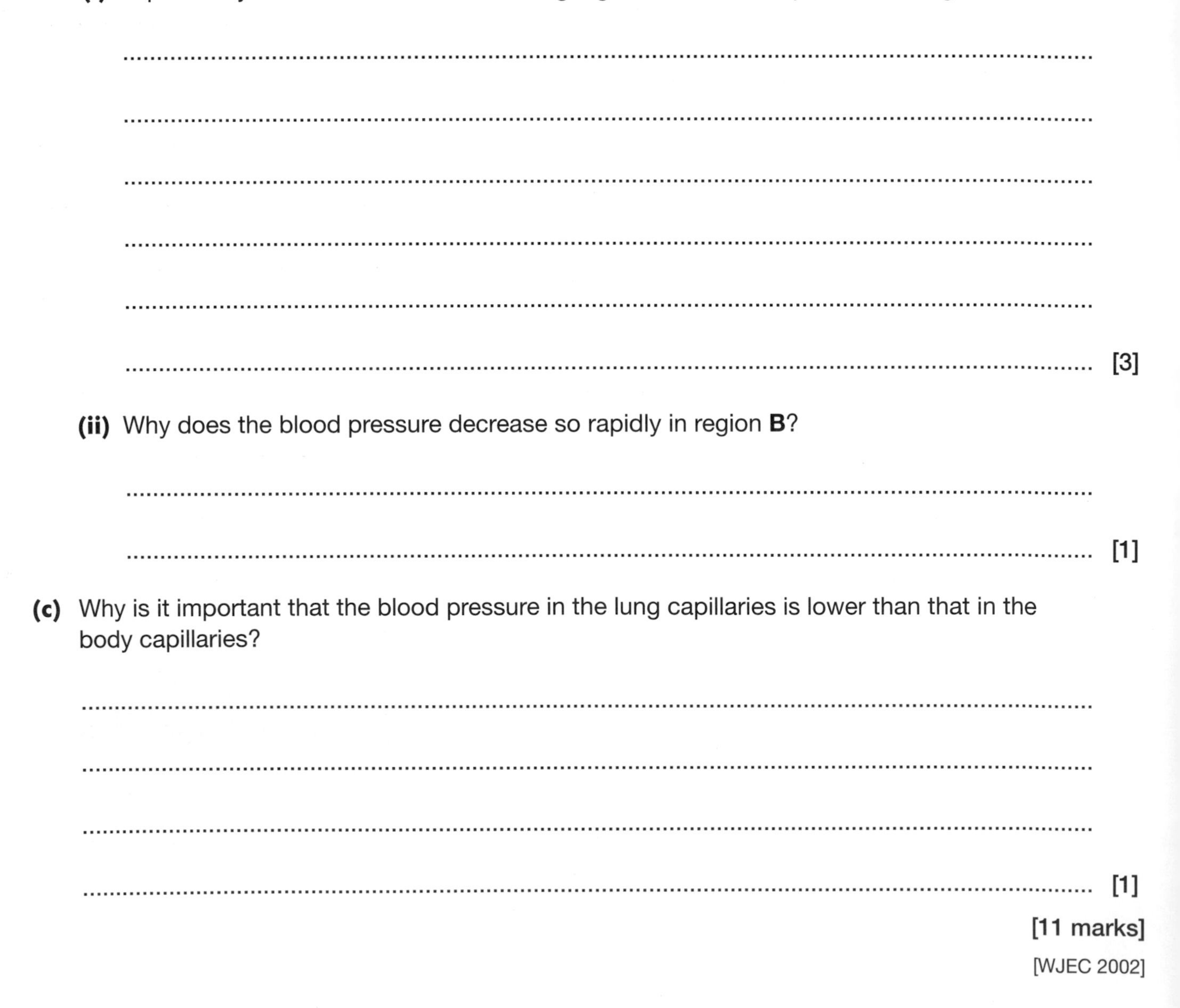

(i) Explain fully the reason for the alternating high and low blood pressure in region **A**.

....

....

....

....

....

.... **[3]**

(ii) Why does the blood pressure decrease so rapidly in region **B**?

....

.... **[1]**

(c) Why is it important that the blood pressure in the lung capillaries is lower than that in the body capillaries?

....

....

....

.... **[1]**

[11 marks]

[WJEC 2002]

Section B

1 The diagrams represent an enzyme and three molecules that could combine with it.

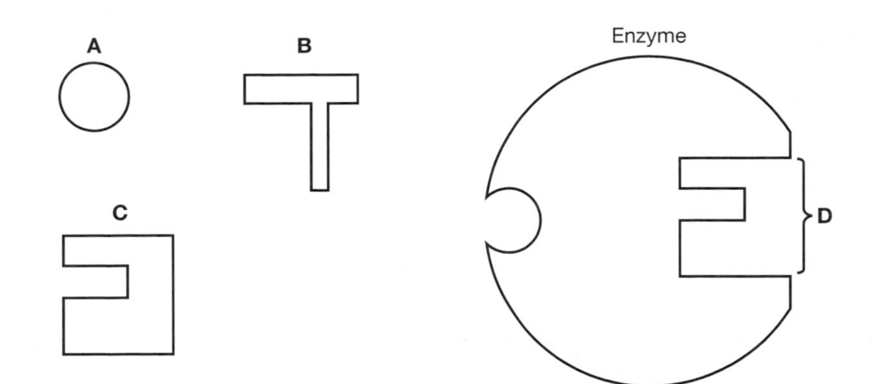

(a) Name the part of the enzyme labelled **D**.

.... **[1]**

(b) Explain how substance **C** is broken down by the enzyme.

....

....

....

....

....

....

....

.... **[4]**

(c) Molecules **A** and **B** inhibit the enzyme in different ways.
Explain how each molecule inhibits the enzyme.

Molecule **A** ....

....

....

....

Molecule **B** ....

....

....

.... [4]

(d) A student carried out an investigation using amylase from a saprophytic fungus. Six wells (holes) of the same size were cut in a starch agar plate. Each well was filled with the same concentration and volume of amylase solution. An equal volume of buffer solution was added to produce a different pH in each well. The plates were incubated at 25 °C for four hours and then covered with iodine solution. It was observed that there were clear rings around each well. The width of these clear rings is shown in the table.

pH	Width of clear ring in mm
4	1
5	2
6	6
7	11
8	9
9	3

(i) Describe how a saprophytic fungus obtains its food.

....

....

....

.... [2]

(ii) What conclusion can be drawn from the results?

....

.... **[1]**

(iii) Use an appropriate method to estimate the maximum rate of reaction that was observed in this investigation. Show your working.

Maximum rate of reaction .... **[2]**

[14 marks]

[AQA B 2002]

2 An experiment was carried out to find the order of twelve amino acids in a polypeptide. The NH_2 and COOH groups at the ends of the polypeptide were radioactively labelled. The polypeptide was hydrolysed to produce shorter fragments. The fragments were separated by paper chromatography. The amino acids in each fragment were identified. The table shows the order of amino acids in each fragment. Each amino acid is represented by a three letter code.

Fragment number	Order of amino acids
1	Val–Tyr
2	Ile–Tyr–Trp–Cys–Asn
3	Tyr–Met–Ile–Tyr
4	Lys–(COOH)
5	Trp–Cys–Asn–Pro–His
6	(H_2N)–Asn–Val–Tyr

(a) What is the primary structure of this polypeptide?

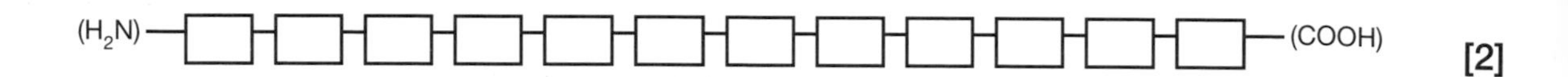

[2]

The diagram shows the chromatogram produced (from the hydrolysed polypeptide).

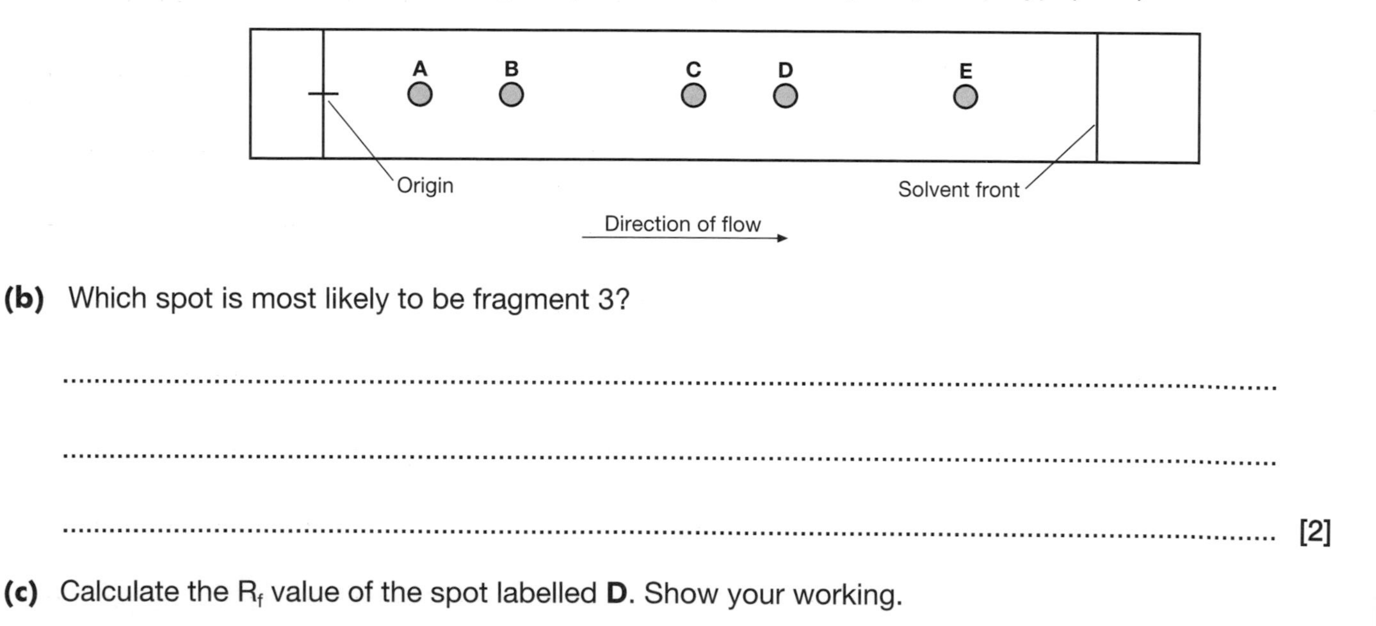

(b) Which spot is most likely to be fragment 3?

....

....

.... [2]

(c) Calculate the R_f value of the spot labelled **D**. Show your working.

R_f value.. [2]

(d) (i) Which spot is likely to consist of more than one type of fragment? Explain your answer.

....

....

....

.... [2]

(ii) Describe how the investigation could modify the experimental technique to separate the fragments completely.

....

....

....

.... [2]

(e) Explain how the structure of the fibrous protein is related to its function.

....

....

....

....

....

....

....

....

....

....

....

.... [5]

[15 marks]

[AQA B 2002]

3 The diagram below shows a section of DNA, which includes a desired gene, that has been cut by a genetic engineer from a human chromosome. This gene codes for a medically important hormone.

desired gene

(a) Explain how this section of DNA would have been cut out of the human chromosome.

....

....

....

....

.... **[2]**

The diagram below shows a bacterial R-plasmid before, and after, the cut section of DNA was inserted into it. The R-plasmid carries genes conferring resistance to the antibiotics, ampicillin and tetracycline.

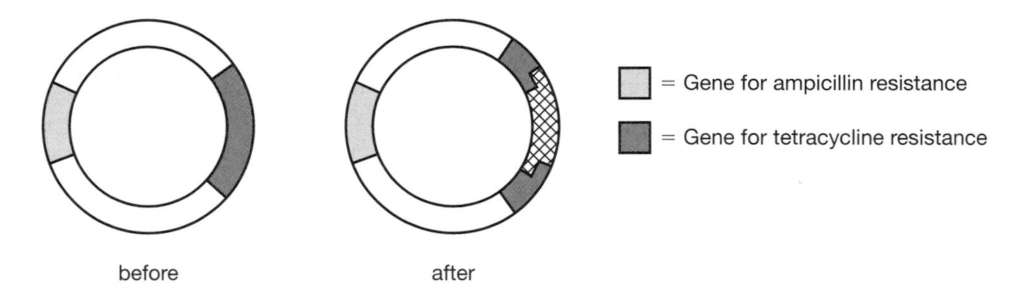

(b) Explain how the section of DNA would have been inserted into the plasmid.

....

....

....

....

.... **[2]**

The plasmids are then mixed with bacterial cells (which do not themselves have plasmids).

(c) Describe how the bacteria may be induced to take up these plasmids.

....

.... **[1]**

Only a few plasmids will have incorporated the sections of DNA. Further, only a few of the host bacteria will have taken up a plasmid of any kind. A genetic engineer needs to determine which bacterial cells possess the plasmids with the inserted section of human DNA before cloning them. This is achieved by culturing bacterial colonies on suitable media.

The bacteria are spread onto a nutrient medium. They are allowed to grow and colonies are then replicated onto plates containing either ampicillin or tetracycline. The bacterial colonies produced on these plates are shown in the diagram below. (The colonies are numbered to allow you to identify them.)

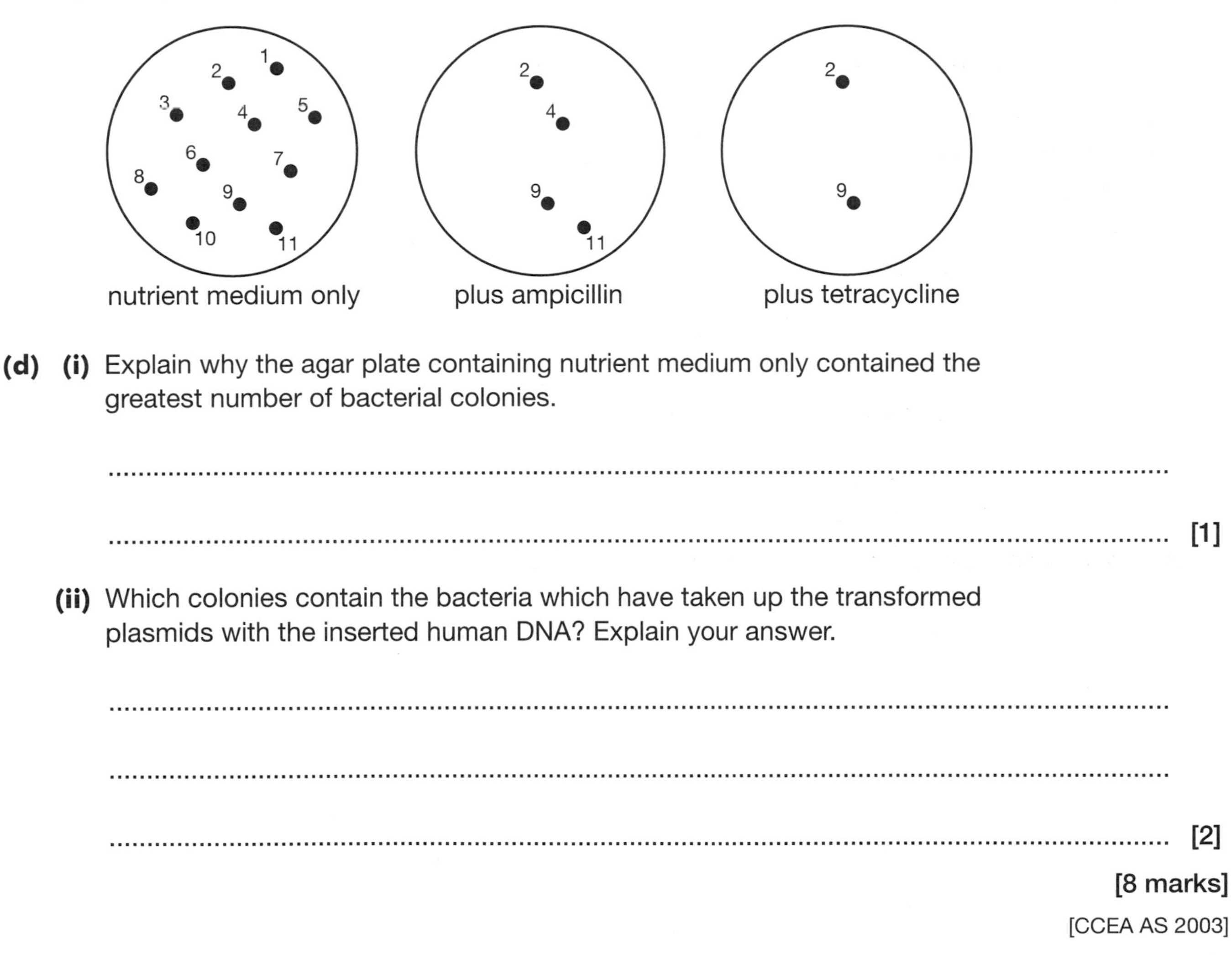

nutrient medium only | plus ampicillin | plus tetracycline

(d) (i) Explain why the agar plate containing nutrient medium only contained the greatest number of bacterial colonies.

....

.... **[1]**

(ii) Which colonies contain the bacteria which have taken up the transformed plasmids with the inserted human DNA? Explain your answer.

....

....

.... **[2]**

[8 marks]

[CCEA AS 2003]

AS Mock Exam 1 Answers

How to use the mark scheme

Symbol	Meaning
;	A separate mark
/	An alternative answer acceptable
max	Mark scheme shows that marks available exceed the question value You can be awarded up to the maximum
underline	When the word or phrase is underlined then it must be given if a mark is to be awarded

Section A

(1) (a) (i) guideline to upper part of atrium;

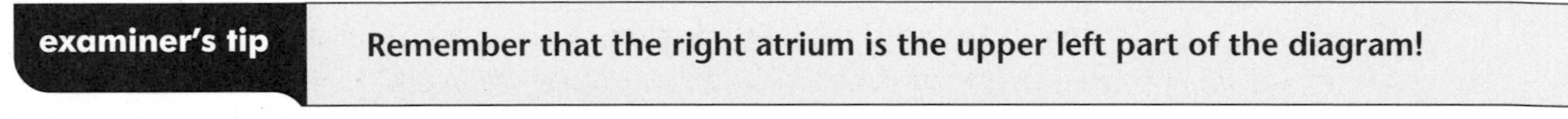
examiner's tip **Remember that the right atrium is the upper left part of the diagram!**

(ii) supply oxygen/nutrients to heart muscle/for *contraction*;

examiner's tip **The Exam Board required that you gave *contraction*. Supply of oxygen alone would not gain credit.**

(b) (i) 0.15 s = 2 marks;
(working 0.04 + 0.10 + 0.01 = 1 mark)
(*allow 1 mark for correct method dividing each distance by rate*)

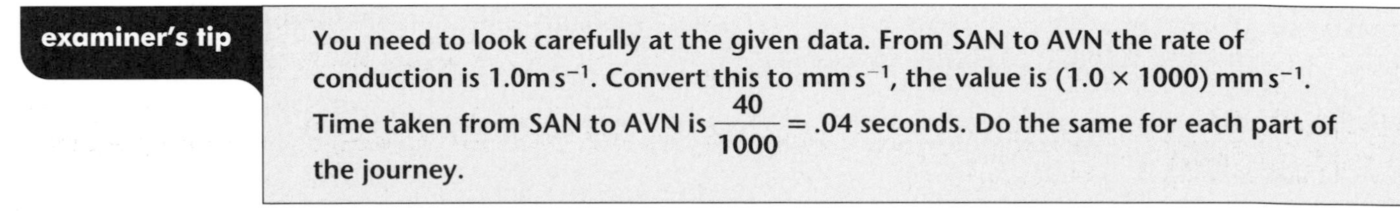
examiner's tip **You need to look carefully at the given data. From SAN to AVN the rate of conduction is 1.0m s^{-1}. Convert this to mm s^{-1}, the value is (1.0 × 1000) mm s^{-1}. Time taken from SAN to AVN is $\frac{40}{1000}$ = .04 seconds. Do the same for each part of the journey.**

(ii) delays contraction of ventricles;
until after atria have contracted/ventricles filled;

examiner's tip **The ventricles need to fill up first before being *stimulated* to contract.**

(iii) rapid contraction of ventricles/both ventricles contract together/rapid transmission to tip so contraction starts at bottom of ventricles; (*reject strong contraction*)

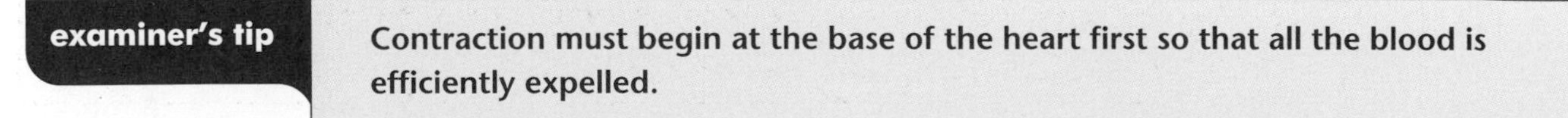
examiner's tip **Contraction must begin at the base of the heart first so that all the blood is efficiently expelled.**

(c) rate of heart beat not adjusted to activity/carries on beating at constant rate/myogenic, so no effect;
(*Accept: rate would increase because parasympathetic normally inhibits*)

examiner's tip **The parasympathetic nerve speeds up the heart and the vagus nerve slows it down when connected to SAN. With no connection you would have the same continuous heart rate, without change.**

(2) (a) Allow any value between the range (41–42) °C;

(b) 46.5 °C;

(c) **(i)** The immobilised form of the enzyme is B;

(ii) 1 The immobilisation process sticks the enzyme to an inert substance. Where the enzyme is attached the substrate molecules are impeded from entering the active site of some enzyme molecules so the activity is lower than for the soluble form.

2 Some enzymes may be denatured by the bonding process so there are less active sites available than with the same concentration of the soluble form of the enzyme.
(1 and 2 in either order are acceptable.)

examiner's tip

Note that if an enzyme is to successfully catalyse a reaction then the substrate and enzyme must collide in such a way as to produce an enzyme–substrate complex. A number of different methods can be used to immobilise enzymes. Some of the enzymes can be denatured by the bonding process. Additionally, access of some substrate molecules to the active site of some of the enzyme molecules can be reduced. Even though the method is efficient from the view of retaining the enzymes to re-use, and having no contamination of product, the extent of reaction is lower!

(d) At higher temperatures the shape of the active site becomes increasingly less suitable for the formation of the enzyme–substrate complex and finally the enzyme denatures.

examiner's tip

Once denatured the enzymes never work again. Note that after the optimum temperature has been reached then the extent of reaction slows increasingly until it reaches zero.

(3) (a) **(i)** A ring around 45 500; **(ii)** $\frac{45\,500}{1\,880\,000}$; Efficiency = 2.42% *(Accept 2.4)*

examiner's tip

The mark scheme gives a mark for the division shown. This needs to be multiplied by 100 to create the percentage 2.4.

(b) Not all light/not all solar energy/is absorbed in photosynthesis; Some energy is dissipated as heat; Some light is reflected; Some light misses the leaves/some light misses the chloroplasts; Overlapping leaves/some leaves shaded; Other named factor may be limiting; (Some) trees are not in leaf all year round; Enzymes are not 100% efficient; *(max 4)*

examiner's tip

'Suggest' always means that your good ideas will count. Stating the 'obvious' is usually beneficial. Here, it is clear that not all light actually hits the tree! Some is reflected, and some heats up objects. So dissipation as heat energy would be given credit.

(c) Heat/thermal;

(4) (a) Protein/polypeptide;

(b) Rough endoplasmic reticulum has ribosomes; This is the site of protein synthesis;

The stem of the question gave you a clue. Rough endoplasmic reticulum has ribosomes and is involved in the assembly of polypeptides from amino acids.

(c) Proteins/polypeptides move to the Golgi apparatus; Reference to protein modification; (Protein or polypeptides) enclosed in membranes to form vesicles; so most activity is in the vesicles after 45 minutes; *(max 3)*

examiner's tip

Follow the maximum radioactivity through each stage.
RER → Golgi apparatus → Secretory vesicles
The sequence should remind you of protein synthesis.

(d) Some amino acids are moving between sites; Amino acids are being broken down/metabolised; Proteins are used in other parts of cells; Proteins are also synthesised in the mitochondria; *(max 2)*

examiner's tip

If the radioactive amino acids did not equate to 100%, then this indicates that 5% were elsewhere in the cell. Did you know that they are broken down in respiration to release energy?

(e) Exocytosis; Vesicles move to the cell membrane; Vesicle fuses with the cell membrane; Contents are released outside of the cell; *(max 3)*

examiner's tip

Clearly if the radioactivity is lost to the cell, then secretion of a product made from radioactive amino acids has taken place.

Section B

(1) *(Answers to part* **(e)** *of this question require continuous prose. Quality of written communication should be considered in crediting points in the mark scheme. In order to gain credit, answers must be expressed logically in clear scientific terms.)*

(a) Blood grouping relies on the presence of agglutinogens/antigens; on the cell surface membrane/plasma membrane of the red blood cells; DNA is found in the nuclei; Only white blood cells have nuclei; *(max 3)*

examiner's tip

You needed to recall that red cells are enucleate (they do not have a nucleus). The early technique relied on the antigen on the cell surface membrane to determine blood group. White blood cells have nuclei which can be used in more modern and accurate techniques.

(b) (i) AB; **(ii)** This showed that the blood has both antigen A and antigen B;

examiner's tip

The presence of the anti-A antibody results in agglutination of red blood cells which have antigen A in their cell surface membrane. Agglutination is the clumping together of red blood cells. In the presence of antibody B the blood also agglutinates. The blood group must be AB!

(c) DNA of which the function is not known/does not code for a protein;

examiner's tip

Vast amounts of DNA have no known function.

(d) (i) Restriction endonuclease/restriction enzyme;

(ii) Restriction enzyme cuts at specific base sequences; isolates non-coding DNA; sequence of bases repeated different numbers of times; number of bases in sequence determines the length of the piece of DNA; *(max 3)*

examiner's tip

Remember that each restriction endonuclease cuts only at highly specific sequences along DNA. Since the base sequence along the DNA of different individuals differs, then the number of times a sequence occurs will differ.

(e) Chains of DNA are separated; by heating; probe consists of complementary sequence; of DNA bases; radioactive; can be located as produces a shadow on photographic plate; *(max 5)*

examiner's tip

A DNA probe is a single DNA strand having a sequence of exposed bases. These bases are ready to bind with complementary bases along the 'suspect' fragments of DNA. T binds with A, and G with C. If the genetic fingerprint is confirmed then the probe and suspect strand bind perfectly. Since the probe has radioactive bases they readily show up on photographic film.

(2) (a) To prevent the entry/exit of water;
by osmosis;
which may affect the organelles; *(Any 2)*

examiner's tip

If the concentration inside the organelle is the same as the concentration of the fluid outside the organelle then there is no net water loss or gain!

(b)

Pellet	Organelle
A	nucleus;
B	mitochondria;
C	ribosomes;

examiner's tip

Remember that in ultracentrifugation the most dense organelles descend to form the first pellet. Once the first pellet has been removed then the supernatant is spun faster and longer which enables the next dense to descend to the bottom of the tube; in this case the mitochondria.
So the ribosomes are in the following pellet, being the next most dense.

(c) glucose/monosaccharides/disaccharides/sugar;
glycogen;
proteins/(poly)peptides;
enzymes (suitable named example);
ions (suitable named example);
amino acids (suitable named example);
lipids/ phospholipids/triglycerides/fats;
microtubules/centrioles;
microfilaments;
RNA/mRNA/tRNA;
(other substances in liver cells are acceptable) *(Any 2)*

examiner's tip

In the fluid above the final pellet of ribosomes there are many components! Take care with this type of question. Never repeat an equivalent substance. Here, if you had given monosaccharides and disaccharides you would have scored one point only. This is because they are both carbohydrates.

(d) Two membranes shown;
inner membrane shown folded;
membrane/envelope;
inter-membrane space;
cristae;
matrix;
ribosomes/DNA/in matrix;
stalked particles/ATPase; *(Any 4)*

examiner's tip

Any 4 of these parts shown in your diagram would score the marks. You do not have to be an excellent artist!

(e) To produce (large amounts of) ATP;
by aerobic respiration/Kreb's cycle/electron transport chain;
(because) liver cells are (metabolically) very active; *(Any 2)*

examiner's tip

Note that the mitochondria are the organelles which enable the release of energy. This would not have been credited at this level. Examiners were looking for knowledge of ATP being produced and the fact that aerobic respiration takes place. More detail is needed at this level.

(3) (a) Translation is the conversion of the base sequence on mRNA into the amino acid sequence of protein.; It is carried out in the ribosomes.; Each ribosome has a small sub-unit and a large sub-unit.; The small unit has a region which will attach to codons of mRNA.; A codon is a sequence of three bases in mRNA coding for one amino acid.; An amino acid is activated; and is attached to a specific tRNA molecule.; This carries the amino acid at one end and a specific anticodon at the other.; The tRNA fits into the large sub-unit.; There are two adjacent sites in each of the sub-units.; As the ribosome passes along the mRNA, one codon at a time, tRNA with the appropriate anticodon fills the vacant slot.; The amino acid forms a peptide bond; with the developing peptide chain held by the tRNA in the adjacent slot.; This continues until a stop codon is reached.; *(max 10)*

examiner's tip

The first thing to make sure of in an essay style response is the detail. Note the number of marks in the scheme, 14, in this instance. Follow this up with good grammar.

(b) Each chromosome consists of two chromatids; Homologous chromosomes consist of one from each parent; and these pair up; The chromatids entwine forming chiasmata; at which point DNA may break and interchange/cross over; In animal cells, centrioles move to opposite poles; Spindle fibres form; In metaphase bivalents become attached to equator; by centromeres; Anaphase – chromosomes of each bivalent pulled to opposite poles; New spindles form at right angles to old; Centromeres divide (kinetochores) and attach to new spindles; Separated chromatids (now chromosomes) are pulled apart; Spindle and chromosomes disappear, nuclear envelope reappears;

(max 10)

AS Mock Exam 2 Answers

How to use the mark scheme

Symbol	Meaning
;	A separate mark
/	An alternative answer acceptable
max	Mark scheme shows that marks available exceed the question value You can be awarded up to the maximum
underline	When the word or phrase is underlined then it must be given if a mark is to be awarded

Section A

(1) (a) light microscope;

examiner's tip

Here you were given quite a problem. Just two basic microscopes to choose from, electron or light microscope? The key evidence is under the photomicrograph, ×1000 the magnification. This is the view you may expect to see under high power in school or college.

(b) **X** = cilia/brush border;
Y = basement membrane;
(not cell or plasma membrane)

examiner's tip

When analysing a tissue like this you need to take care to identify X correctly. The cell membrane on the dorsal surface has tiny, thin 'fibre-like' parts. They are cilia but you may be tempted to write microvilli. This would be wrong!
Y, the basement membrane, is a structure that the cells adhere to.

(c) Cilia beat/move/wave;
Mucus carried along/mucus moved/mucus removed;
(not reference to goblet cells)

examiner's tip

Note that the examiners did not credit reference to goblet cells. They produce mucus, but it is the cilia which move this mucus along, the cilia move backwards and forwards like barley in a corn field. The mucus is moved along.

(d) Distance allowed 0.04–0.0425 mm

$\frac{33}{800}$; 0.04125 mm or 41.25 μm;

examiner's tip

Look out for this type of question in the future. You needed to measure the height of the cell, which is any value between the extremes of 34–36 mm. The photomicrograph is magnified ×1000 so you divide by this figure to find the true height.

(e) Collection/many/(large) number/of cells/group (not layer);
Similar structure;
Carrying out a particular function/working together; (Any 2)

examiner's tip

A **collection** of **similar** cells **working together** on a **specific function**. This is typical of material you must revise carefully. Being nearly correct does not score marks. Note the key words in the definition. Try to revise like this!

(2) (a) (i) A = Carbohydrate/glycocalyx/glycoprotein/polysaccharide;
(not monosaccharide/sugars)
B = Phosphate/polar group/hydrophilic head/phospholipid head;
C = Hydrocarbon/non-polar heads/hydrophobic tails/fatty acids;
(**not** tails/not lipid layer)
D = Transmembrane carrier/carrier protein/channel protein/intrinsic protein;
(**not** protein)
E = Protein/extrinsic protein;

examiner's tip

Do not be worried when you see an extensive mark scheme like this, with many possible answers. Remember that you only need one alternative for each! A consists of a linear string of sugar units shown by the hexagonal shapes. The most logical answer is carbohydrate.
The mark scheme does show the detail needed. Giving protein for part D would not score a mark even though it is a protein! The qualifying detail needed was carrier protein or equivalent. Detail is a requirement at this level and this builds to a greater need for additional detail in A2.

(ii) Fluid mosaic/Singer and Nicholson;

examiner's tip

The model shown in the diagram was first suggested by Singer and Nicholson, hence their names achieve credit. They gave the term fluid mosaic as a description.

(b) Glucose
Polar molecule;
Phospholipid layer/hydrophobic region impermeable;
Protein channels provide passage;
Hydrophilic linings to channels; allow water-filled pores
Ref. protein carriers/transporter molecules/transmembrane protein;
Ref. facilitated diffusion; *(Any 2)*
Vitamin A
Non-polar;
Passes through directly/no channels required;
Dissolves in phospholipid layer/dissolves in hydrophobic regions;
Diffusion; *(Any 2)*

examiner's tip

This question finds out if you know how a water soluble substance (glucose) can cross the plasma membrane, and the different way that a fat soluble substance can cross. Glucose cannot get across the phospholipid parts so must have a carrier protein to 'make the crossing'.
Vitamin A can cross the phospholipid bilayer as it is fat soluble. The vitamin A can dissolve in the phospholipids and diffuse across directly!

(c) Active transport;
Osmosis;
Endocytosis/exocytosis/phagocytosis/pinocytosis or suitable description; *(Any 2)*
(not ref. to diffusion or facilitated diffusion)

examiner's tip

The first two alternatives should be easy to recall. If you chose the third alternative there is a danger. The examiners considered that endocytosis and exocytosis were equivalent marking points, even though their transport directions were opposite. Ensuring that the alternatives you give are very different and unrelated is usually a good policy.

(3) (a) (i) Metaphase;

examiner's tip

Tricky view! Look at the broken line which confirms that all of the bivalents are at the equator. This pinpoints the stage as metaphase.

(ii) Centromeres divide;
Chromatids separate/pulled apart;
by spindle fibres; *(Any 2)*

examiner's tip

The spindle fibres contract and pull the centromere apart. A common error is to state that the chromosomes are separated. Wrong! They are known as chromatids until they are parted.

(iii) Three chromosomes;
One of each homologous pair;

examiner's tip

Danger! Up to now this question has been about mitosis but in part (iii) the examiners have switched to meiosis. (Tricky stuff!) Look at the diagram which shows that the cell's diploid number of chromosomes would be 6, so the gamete number is half of this, i.e. 3. Draw the chromosome to an accurate size. The correct shading would also help.

(b) 7.6 is replicated DNA/chromatids joined together/
late interphase/prophase/metaphase/before cell division;
3.8 contains single chromatids/DNA has not replicated/
telophase/early interphase;

examiner's tip

The cells are clearly at various stages of cell division. The 7.6 units of DNA represent the amount after DNA replication has taken place. This amount is finally halved after the chromatids have been separated as a result of anaphase.

(4) (a)

Vein	Large diameter thin walls; Valves;	Easily compressed by muscle/decreased resistance to flow of blood; Prevents backflow of blood;
Capillaries	One cell thick; Cells have gaps in between them;	Minimum/easier distance for diffusion/exchange; Allows tissue fluid to escape from vessel; (not pressure difference due to diameter)
Arteries	Thick muscle/layer of elastic fibres/ elastin;	Can withstand high blood pressure/allows elastic recoil of vessel/maintains high blood pressure;
Allow this once only for any blood vessel	Smooth endothelium;	Minimises friction of blood on vessel wall;

Two marks for each pair of characteristic + reasoning for adaptation.
One mark if all characteristics given but no matched reason.

examiner's tip

Veins have a large diameter which gives less resistance to blood flow. This is important because the veins have little muscle (so little contraction!) and are furthest from the heart. Low resistance enables the veins to make best use of the lower pressure to enable progress of blood towards the heart. It must be easy for the skeletal muscles to compress veins to push the blood towards the heart.
Capillaries consist of squamous epithelium which have a short diffusion path to exchange chemicals.
Arteries have a thick layer of smooth muscle which contracts to propel blood. This, and elastin, enables the vessel to withstand higher blood pressure.

(b) (i) blood enters aorta from left ventricle;
left ventricle contracts so gives peaks in region A;
left ventricle relaxes gives troughs in region A;
aorta itself contracts; *(max 3)*

(ii) arterioles dilated (so blood moves through easily);

(c) if pressure in body tissues was lower the blood would tend to flow in wrong direction/makes sure that the pressure gradient is in right direction/;

Section B

(1) (a) Active site;

(b) Substrate enters active site;
Substrate and active site are complementary shapes/lock and key;
(binding) to form enzyme–substrate complex;
Lowering of activation energy;
Conformational change/shape change;
Breaking of bonds in substrate;
Products no longer fit the active site so are released; *(Any 4)*

examiner's tip

Again, this type of answer needs detail. Most candidates would score the first mark, but other details are higher concepts. Underlining in the mark scheme means that those words must be given. Learn this sequence ready for your examination. It is a question regularly asked.

(c) Molecule A binds at a site away from the active site/allosteric site;
Causes enzyme/active site to change shape;

Molecule B can enter/competes for active site;
Prevents substrate from entering/no enzyme substrate–complex formed/active site blocked;

examiner's tip

Molecule A is a non-competitive inhibitor which clearly fits into the alternative site. It is an allosteric site and not another active site. Once it binds to the allosteric site then the shape of the active site is changed so that it is not an ideal shape in which the substrate can fit.
Molecule B is a competitive inhibitor which clearly fits into the active site. It is not the same shape as the substrate but it still can bind inside of the active site to prevent the substrate entry.

(d) (i) Secretes enzymes (for extra-cellular digestion);
Absorbs products;

examiner's tip

Note that saprophytes obtain their nutrients by secreting extracellular enzymes into dead organic material. The enzymes break down the external substrates into soluble products which are then absorbed. Note the underlined words which are required for credit.

(ii) Optimum pH is 7/neutral/between 6 and 8/between 7 and 8;

examiner's tip

The starch-agar plates were flooded with iodine solution so that where there was still some starch it would show up as blue-black. A clear ring shows where the amylase has changed the starch to maltose. The biggest clear ring was at pH 7 so that was the optimum pH for the fungal amylase.

(iii) Maximum rate $= \frac{\text{Distance}}{\text{Time}} / \frac{11}{4} / \frac{11}{4 \times 60}$;
[Correct answer = 2 marks (ignore units)
i.e. 2.75 mm/hour, 0.046 mm/min, 4.6×10^{-3} mm/min/
1 mm/21.8 min, 23.76 mm^2/hour]

examiner's tip

Never fear maths in biology papers! Here you need to focus on the rate of reaction. The examiners have given you a range of acceptable alternatives to calculate the rate of reaction. Rate always involves the division of time into the process.
11 mm/4 hours = 2.75 mm/h
$11/4 \times 60 = 0.046$ mm/min or 4.6×10^{-3} mm/min
21.8 minutes for each 1 mm increase
Clear ring is 5.5 mm radius, so area of ring $\frac{3.142 \times 5.5 \times 5.5}{4} \frac{\text{mm}^2}{\text{hours}} = 23.76$ mm^2/hour

(2) (a) Correct sequence = 1;

(H$_2$N)–Asn–Val–Tyr–Met–Ile–Tyr–Trp–Cys–Asn–Pro–His–Lys–(COOH)

Both correct = 1;

examiner's tip

With a question like this it is vital to spot the key information. You are informed that there is a polypeptide consisting of 12 amino acids. However there are 20 in the table. The process of hydrolysis has cut the polypeptide at several different positions forming fragments of different lengths. You are shown the position of H$_2$N in the given structure and COOH at the end, so fragments 6 and 4 from the table go at the beginning and end respectively, as above.
Now for the middle sequence! The principle to work this out must be that the beginning of one fragment must be the same as the end of another, so link the following fragments in this order,
6→1→3→2→5→4. Write out all of the amino acid sequences of these fragments in a line, then cross out any parts of the sequence which are repeated! See below

(H$_2$N) – Asn –Val –Tyr –Val –Tyr –Tyr – Met – Ile –Tyr – Ile –Tyr –Trp – Cys – Asn –Trp – Cys – Asn – Pro – His – Lys – (COOH)

Each amino acid underlined should be crossed out!

(b) B; bigger molecules move more slowly. Fragment B has 4 amino acids, second biggest fragment;

(c) R_f = Distance travelled by the spot/Distance travelled by solvent front;
62 mm/102 mm = 0.61;
Must be 2 decimal places
Correct answer = 2 marks

examiner's tip

If you show no working and simply do the calculation on your calculator then you score two marks. Always show your working; if you have not given the correct answer the examiner looks back and can give credit for the method.

(d) (i) A;
As there are two molecules composed of 5 amino acids/
As there are two molecules the same size/
it contains fragments 2 and 5;

examiner's tip

The longer chains of amino acids move along the chromatography paper more slowly so they must be nearer the start line. There are 2 long fragments exactly the same size, i.e. 5 amino acids.

(ii) Use 2 way/2D chromatography/rotate chromatogram 90°;
Use a different solvent;

examiner's tip

Always remember that turning the chromatogram at right angles may enable 2 substances at a spot to be further separated.

(e) Long chains of amino acids;
Folding of chain into a coil/folds/helix/pleated sheet;
Association of several polypeptide chains together;
Formation of fibres/sheets explained;
H bonds/disulphide bonding (in context);
Fibres provide strength and (flexibility);
Sheets provide flexibility;
Examples such as keratin in air/collagen in bone; (must be in context)
Insoluble because external R-groups are non polar;

examiner's tip

For this type of question you need to know structures 1–4 of proteins. Knowledge of this is regularly tested across the Examination Groups. Explaining the primary, secondary, tertiary and quaternary structures scores 4 of the marks above. Examples are regularly given credit so it is worthwhile learning several.

(3) (a) Use of restriction endonuclease;
Cuts the DNA at specific sequence;
leaves sticky ends at each end of DNA/cuts in DNA strands are staggered; *(Any 2)*

examiner's tip

Some Exam Groups would credit restriction enzyme but it is not recommended that you risk this. Always give restriction endonuclease and be aware that there are many different endonucleases. Often students give reverse transcriptase in answer to this type of question. This is wrong!

(b) Cut R-plasmid using the same restriction endonuclase;
sticky ends at either end of human DNA bind to complementary ends of plasmid;
ends joined/annealed by ligase; *(Any 2)*

examiner's tip

If the same restriction endonuclease was not used then the sequences at the sticky ends would not be complementary. The ligase enzyme allows the desired gene to be joined in the plasmid. Without the ligase the new gene would remain isolated!

(c) Giving bacterial cells a brief electrical shock/electroporation;

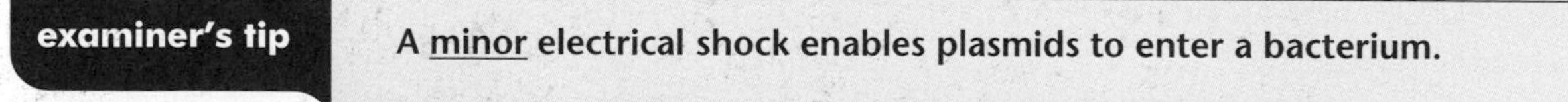
examiner's tip

A minor electrical shock enables plasmids to enter a bacterium.

(d) (i) The plate with agar only does not have any antibiotic to kill the bacteria;

examiner's tip

The plates with nutrient alone have enough nutrient to support bacterial growth but no antibiotic, so all bacteria live, and grow into colonies.

(ii) Colonies 4 and 11;
They have the desired gene inserted into the middle of gene for tetracycline resistance which is not now effective;

examiner's tip

The key diagrams show the plasmid before and after insertion of the desired gene. This gene is inserted into the middle of the tetracycline resistance gene on the plasmid which no longer has the ability to code for tetracycline resistance.